民航空中交通管理系列丛书

空中交通管制员无线电陆空通话

（第 2 版）

王万乐　赵　琦　编著

清华大学出版社

北京交通大学出版社

·北京·

内 容 简 介

本书是民航院校空中交通管理（以下简称空管）专业学生学习掌握基础无线电陆空通话的教材，内容主要按照飞行阶段组织，包括基本运行程序、机场管制、进近管制、区域管制和紧急情况，附录提供了国内外常见航空公司代码、话呼及公司名称。对于每个知识点，参考国际民航组织及我国相关文件，列出了常用单词和词组及英汉双语对照的通话范例。每个知识点由"听录音"引入，符合陆空通话的应用场景。本书配有题型丰富的练习题，辅以图片和音视频，并适当提高练习的挑战度，便于开展教学活动。读者可通过扫描本书扉页上的二维码下载相关录音文件。

本书除用于空管专业教学外，还可作为在职管制员准备中国民航管制员英语等级测试的参考书。

图书在版编目（CIP）数据

空中交通管制员无线电陆空通话 / 王万乐，赵琦编著 . —2 版 . —北京：北京交通大学出版社：清华大学出版社，2022.8

ISBN 978-7-5121-4766-9

Ⅰ．①空…　Ⅱ．①王…　②赵…　Ⅲ．①空中交通管制-陆空协同通信　Ⅳ．①V355.1

中国版本图书馆 CIP 数据核字（2022）第 130738 号

空中交通管制员无线电陆空通话
KONGZHONG JIAOTONG GUANZHIYUAN WUXIANDIAN LUKONG TONGHUA

责任编辑：谭文芳
出版发行：清华大学出版社　　　邮编：100084　　电话：010-62776969　http://www.tup.com.cn
　　　　　北京交通大学出版社　　邮编：100044　　电话：010-51686414　http://www.bjtup.com.cn
印 刷 者：北京时代华都印刷有限公司
经　　销：全国新华书店
开　　本：185mm×260mm　　印张：14.25　　字数：364 千字
版 印 次：2016 年 5 月第 1 版　　2022 年 8 月第 2 版　　2022 年 8 月第 1 次印刷
印　　数：1～2 000 册　　定价：56.00 元

本书如有质量问题，请向北京交通大学出版社质监组反映。对您的意见和批评，我们表示欢迎和感谢。

投诉电话：010-51686043，51686008；传真：010-62225406；E-mail：press@bjtu.edu.cn。

第 2 版前言

自 2016 年本书第 1 版正式出版以来，已经过六年的教学实践检验。实践证明，"听说为主、情境辅助、贴合实际、讲练结合"的教材编写思路受到了广大师生的欢迎。

在体系结构上，本书以航空器飞行阶段为主线安排通话术语。对每个术语知识点，按照简介、学习目标、单词与词组、听录音、结构格式和范例进行呈现，尤其是听录音环节的安排，强调从"听"引入知识点，便于教师引导启发学生发现所听通话的规律和要点，再结合题型丰富的练习题开展听说训练。

在内容上，考虑到陆空通话的特点，本书不以知识呈现为主，而是通过大量例句，将术语融入其中。在内容编排顺序上，通过在例句和练习中提前设计合适的知识点，适当提高教材和练习的挑战度，避免平铺直叙，使学生失去兴趣。此外，结合实际运行中的关注要点（如"相似航班号"）和实际运行情境编排练习，展现陆空通话的应用场景。

第 2 版在保持第 1 版教材原有特色的基础上，主要进行了以下 4 个方面的更新。

（1）根据课程思政建设需求，在每章"本章综合练习"的最后增加"小拓展"活动，提供更多音视频资源和互动练习，培养学生服务意识、责任意识和职业素养，帮助学生提高陆空通话中的沟通交流能力。

（2）根据当前运行环境下陆空通话要求进行内容更新，如新版跑道表面状况全球报告格式（GRF）实施后，跑道表面状况通报术语发生的变化。

（3）根据第 1 版教材的使用反馈，将《空中交通管制员无线电陆空通话》和《空中交通管制员陆空通话练习册》合并为一本教材，以便于使用。

（4）对第 1 版中存在的错误进行更正。

第 2 版主要由王万乐和赵琦修订完成。中国民航大学空中交通管理学院英语课程组教师赵德斌、黄贻刚、李学明、刘博、宋祥波、赵璐、杨越和刘永欣，以及空中交通管理学院各届使用第 1 版教材的学生对本书提出了修改建议，在此表示衷心的感谢。

由于时间和编者水平有限，书中难免出现疏漏之处。我们衷心希望专家、同行和广大读者不吝指正，以便今后继续完善本书。

编 者
2022 年 5 月

第 1 版前言

无线电陆空通话（以下简称陆空通话）是当前空中交通服务单位与航空器之间进行话音通信的方式。正确使用陆空通话用语进行信息沟通，对航空器的安全高效运行有着至关重要的作用。历史上由于陆空通话不规范导致航空不安全事件甚至飞行事故的案例不胜枚举，为此，国际民航组织（ICAO）提出一系列英语语言标准和建议，强调了相关通话英语的重要性，并且在《ICAO 语言能力要求实施手册》（Doc9835/AN/453）中明确了英语语言熟练程度的评判标准。从事国际飞行的航空器驾驶员（以下简称驾驶员）和空中交通管制员（以下简称管制员）须达到相应英语通话标准要求，除了使用 ICAO 规定的标准术语（phraseology），还必须能够用简明英语（plain language）准确、简洁、无误地进行交流。中国民用航空局空中交通管理局要求管制员必须达到 ICAO 所要求的管制员英语等级四级及以上水平。

陆空通话作为驾驶员和管制员在整个飞行过程中按照一定程序进行交流、沟通的语言，具有很强的专业性和情境性。陆空通话教学的根本目的就是提高学生实际应用英语通话的能力，主要是"听"与"说"方面，让学生具备在非面对面的情况下准确实现信息交流的能力。学生从开始接触陆空通话到正确、熟练地应用标准术语以至达到"脱口而出"的程度，需要一个长期的、科学的学习、练习和应用过程。本书针对民航院校空管专业学生，以陆空通话术语为核心，在内容编排上以飞行流程为主线，从"听"入手，形成对陆空通话应用情境的基本认识，再结合实例、示意图等讲述飞行各阶段的标准通话术语，旨在将通话内容的学习与练习置于空管运行情境中，使学生在具体的场景中练习通话、提高能力。此外，本书从 ICAO 语言熟练要求出发，即以发音、结构、词汇、流利程度、理解能力和互动 6 个方面为着眼点设计了大量的听说练习，充分体现"学练结合、以练代学、练中生技"的学习方法和规律，旨在培养学生基本的陆空通话能力和习惯，为将来进入空管岗位进行更高阶段的培训打下坚实的基础。万丈高楼平地起，对于民航院校空管专业的学生来说，掌握基本的陆空通话术语并建立良好的通话习惯是学习期间必须不断努力追求的目标。

本书由中国民航大学空中交通管理学院教师王万乐和黄贻刚提出最初的编写思路和核心理念，王万乐和赵琦主编，英语课程组教师李学明、赵德斌、马丽娟、刘博、赵璐和刘永欣参编，终成此稿。本书共分为 7 章。第 1 章、第 4 章及第 7 章主要由王万乐和黄贻刚编写；第 2 章主要由李学明和赵璐编写；第 3 章主要由赵琦编写；第 5 章主要由赵德斌和刘永欣编写；第 6 章主要由刘博和马丽娟编写。

　　中国民用航空中南地区管理局空中交通管制中心的钱旭、夏卫兵及深圳空管站航务部的艾国胜对本书部分内容的编写提供了悉心指导，保证了本书内容与"一线"实际所用的一致性。本书配套录音由 Todd Kendall（新西兰空中交通管制员、管制教员）和王万乐录制完成。中国民航大学空中交通管理学院学生吉瑞、高雪岩和冯佳怡对本书亦有贡献。在此一并表示衷心的感谢。

　　由于时间和编者水平有限，书中难免出现疏漏之处。我们衷心希望专家、同行和广大读者不吝指正，以完善本书。

编　者
2016 年 4 月

目　录

第1章　基本运行程序

由于无线电陆空通话对航空安全的重要意义及其有别于日常交流的特殊性，国际民航组织对通话语言结构及内容制定了严格的标准。空中交通无线电通话用语是一种半人造语言，总体上具有严谨、简洁、无歧义等特点。

基本运行程序是所有管制员在指挥航空器时需要执行的通用程序，其中包括具有基础性、普遍性等特点的陆空通话。本章主要包括字母、数字、标准单词和词组、呼号及通信的基本方法等内容。

通过本章的学习，应达到以下学习目标：

❖ 掌握陆空通话的基本特点；
❖ 掌握字母、数字在陆空通话中的正确读音；
❖ 掌握标准单词和词组的正确发音及其含义；
❖ 掌握呼号的表达方法，以及常见航空公司三字代码的正确读音，注意区分相似呼号；
❖ 掌握通信的建立程序和一般原则；
❖ 理解正确使用标准陆空通话术语的重要性。

1.1　概述

1.1.1　通话基本要求

陆空通话使用送话器发送，采用按键通话（push-to-talk，PTT）方式。在同一通信频率内，任何时刻只应有一个电台发话，其他电台收听，否则容易出现波道卡阻的现象。

管制员在与驾驶员通话前应检查接收机音量是否设置在最佳音量位置，并确认波道中没有其他通信干扰。

陆空通话中应使用中文（普通话）或英文，时间采用协调世界时（UTC 时间）。

管制员应发出完整且含义清楚的指令，避免让航空器驾驶员无所适从或无法操纵。

为避免歧义或误解，必须严格按照标准陆空通话的要求通话，且应注意以下方面：

（1）先想后说，应在发话之前想好说话内容，保证发话语句完整、流畅；

（2）先听后说，应避免干扰他人通话，确认他人通话完毕之后再发话；

（3）熟练掌握送话器使用技巧；

（4）发话速度适中，在发送需记录的信息时降低速率，使用英文通话时，每分钟发话不超过 100 个单词；

（5）发音清楚，保持通话音量平稳，语调正常；

（6）在通话中，数字前应稍作停顿，重读数字并以较慢语速发出，以便于理解；

（7）应避免使用"啊""哦"等犹豫不决的词；

（8）应在开始通话前按下发送开关，待发话完毕后再将其松开，以保证通话内容的完整性；

（9）应注意发话的语音、语调和节奏；

（10）一般情况下，每次发送的管制指令不宜包含过多内容，且应将相对重要的内容排列在管制指令的尾部。

1.1.2　通话结构

首次联系时应采用的通话结构为：

对方呼号+己方呼号+通话内容

例如：

> **P:** Dongfang Approach, CCA1356, 5700 meters maintaining.
> **C:** CCA1356, Dongfang Approach, radar contact.
>
> **P:** 东方进近，CCA1356，五拐保持[①]。
> **C:** CCA1356，东方进近，雷达看到。

首次通话以后的各次通话，管制员采用的通话结构一般为：

对方呼号+通话内容

例如：

> **C:** CES3662, climb to 6000 meters.
> **P:** Climbing to 6000 meters, CES3662.
>
> **C:** CES3362，上升到六千。
> **P:** 上升到六千，CES3662。

驾驶员采用的通话结构为：

对方呼号+己方呼号+通话内容

例如：

> **P:** Dongfang Tower, CSN6583, request push-back and start up.
> **C:** CSN6583, push-back and start up approved.
>
> **P:** 东方塔台，CSN6583，请求推出开车。
> **C:** CSN6583，同意推出开车。

管制员确认驾驶员复诵的内容时可仅呼对方呼号。当管制员认为有必要时，可采用"read-back correct"具体确认。例如：

> **C:** CDG1152, taxi to holding point Runway 34 via Taxiways A, A7, Y and Z1.

---

① 为了强调高度指令的中文读法，本章涉及高度的中文指令均以读法显示。其他章节用阿拉伯数字表示的高度信息，其中文读法请参考 1.2 节和 1.3 节。

P: Taxi to holding point Runway 34 via Taxiways A, A7, Y and Z1, CDG1152.

C: CDG1152, read-back correct.

C: CDG1152，沿滑行道 A，A7，Y 和 Z1 滑到 34 号跑道等待点。

P: 沿滑行道 A，A7，Y 和 Z1 滑到 34 号跑道等待点，CDG1152。

C: CDG1152，复诵正确。

练 习 题

1. 听说练习

说明：听录音并跟读。

（1）**P:** Dongfang Approach, CHH7345.

　　　C: CHH7345, Dongfang Approach.

（2）**C:** CQH8814, Beijing Approach, climb to and maintain 5400 meters.

　　　P: Climb to and maintain 5400 meters, CQH8814.

（3）**P:** Beijing Control, LKE9988, leaving 7800 meters descending to 7200 meters.

　　　C: LKE9988, roger.

（4）**C:** KLM898, Dongfang Tower, are you ready for departure?

　　　P: Ready, KLM898.

（5）**C:** CXA8130, Dongfang Control, confirm squawk 2343.

　　　P: Affirm, CXA8130.

2. 完成通话

说明：你作为一名管制员，有一架航空器与你建立首次联系，之后你向该航空器发布高度指令。请从下框中选取适当内容组成正确的通话。

| BAW038　descend to 4500 meters　climb to 5400 meters　CES2105　CCA1108 |
| climb to 6300 meters　descend to 9500 meters　Dongfang Control　Dongfang Approach |

（1）**P:**_____.

　　　C:_____.

　　　C:_____.

　　　P:_____.

（2）**P:**_____.

　　　C:_____.

　　　C:_____.

　　　P:_____.

3. 听说练习

说明：结对活动。首先，学生 A 读 Student A 的内容，学生 B 听并记录。在听并记录的过程中，如果不确定，学生 B 可以采用 "say again" 来澄清。然后，两人交换角色。练习后可跟读录音。

利用"say again"澄清的方法如表 1-1 所示。

表 1-1

用　　语	含　　义
say again（重复）	重复完整信息
say again...（item）（重复……）	重复特定信息项
say again all before...（重复……之前）	重复第一个接收清楚单词之前的那部分信息
say again all after...（重复……之后）	重复最后一个接收清楚单词之后的那部分信息
say again all between...and...（重复……和……之间）	重复在两个接收清楚单词之间的那部分信息

Student A

（1）CES1202, line up, be ready for immediate departure.

（2）Dongfang Control, LKE9988, DGT at 30, maintaining 10100 meters, estimating BRAIT at 50.

（3）CHH7345, turn left heading 330, cleared for ILS approach, Runway 36, report established.

（4）Dongfang Tower, DLH723, request push-back and start up, Stand 15, Information C.

Student B

（1）GCR6911, Tianjin Approach, descend to 4500 meters on QNH 1001.

（2）CSN3369, Dongfang Tower, Runway 06, cleared for take-off.

（3）CCA1389, Dongfang Control, unknown traffic, 10 o'clock, 15 kilometers, crossing from left to right.

（4）DLH723, Dongfang Approach, position 20 kilometers east of the field.

1.2　字母

在无线电通话中，管制员和驾驶员都使用国际统一的标准字母，通过用单词为英文字母注音的方法来避免发音混淆。

1.2.1　听录音 🔊

表 1-2 为字母的标准读法，发音一栏中带下划线的部分应重读。请听录音并跟读。

表 1-2

字　　母	单　　词	发　　音
A	Alpha	<u>AL</u> FAH
B	Bravo	<u>BRAH</u> VOH
C	Charlie	<u>CHAR</u> LEE or <u>SHAR</u> LEE
D	Delta	<u>DELL</u> TAH

续表

字　母	单　词	发　音
E	Echo	ECK OH
F	Foxtrot	FOKS TROT
G	Golf	GOLF
H	Hotel	HO TELL
I	India	IN DEE AH
J	Juliett	JEW LEE ETT
K	Kilo	KEY LOH
L	Lima	LEE MAH
M	Mike	MIKE
N	November	NO VEM BER
O	Oscar	OSS CAH
P	Papa	PAH PAH
Q	Quebec	KEH BECK
R	Romeo	ROW ME OH
S	Sierra	SEE AIR RAH
T	Tango	TANG GO
U	Uniform	YOU NEE FORM or OO NEE FORM
V	Victor	VIK TAH
W	Whiskey	WISS KEY
X	X-ray	ECKS RAY
Y	Yankee	YANG KEY
Z	Zulu	ZOO LOO

通话中，字母或字母组合可用来表示机场识别代码、信标台名称和航路点等，且应按无线电发音方式的发音规则读出。但需要注意的是，有些约定俗成的字母组合读法与此不同，如 ATIS、QNH、VOR、NDB 等，有的可按照一个单词读出，而有的则只需按照英文字母表的字母读法逐位读出。

1.2.2　机场识别代码

机场识别代码按无线电发音方式逐位读出，如表 1-3 所示。

表 1-3

机场识别代码	英 文 读 法	中 文 读 法
RJTY	ROMEO JULIETT TANGO YANKEE	ROMEO JULIETT TANGO YANKEE
ZBAA	ZULU BRAVO ALPHA ALPHA	ZULU BRAVO ALPHA ALPHA
EDDF	ECHO DELTA DELTA FOXTROT	ECHO DELTA DELTA FOXTROT

1.2.3 全向信标（VOR）台和无方向性信标（NDB）台

英文读法为按照字母无线电发音读出该台识别码，中文读法为按照航图中的地名读出，如表 1-4 所示。

表 1-4

VOR 和 NDB	英 文 读 法	中 文 读 法
SIA	SIERRA INDIA ALPHA	西安
VYK	VICTOR YANKEE KILO	大王庄
SX	SIERRA X-RAY	南浔

当 VOR 和 NDB 导航台名称相同、不建在一起且距离较远时，中文读法应在台名后加 VOR 或 NDB，如表 1-5 所示。

表 1-5

VOR 和 NDB	英 文 读 法	中 文 读 法
POU	PAPA OSCAR UNIFORM	平洲 VOR
XK	X-RAY KILO	平洲 NDB

1.2.4 航路点

若航路点是五个英文字母，则中英文读法相同，约定俗成按照一个单词的英文发音读出。若航路点是由 P 和数字组成，英文按照无线电字母发音"PAPA"加数字的英文发音读出，中文则按照字母"P"加数字的中文发音读出，如表 1-6 所示。

表 1-6

航 路 点	英 文 读 法	中 文 读 法
ANDIN	ANDIN	ANDIN
EPGAM	EPGAM	EPGAM
P23	PAPA TOO TREE	P 两三

1.2.5 航路

航路由航路代号和编码组成，分别按照数字和字母的发音读出。航路代号前有"U""K""S"时，分别读作"upper""kopter""supersonic"，表示高空、直升机、超音速航路。标准进离场航线的英文读法按照字母和数字的发音，后加"arrival"或"departure"读出，中文读法为导航台名称+数字和字母的发音，后加"进场"或"离场"，如表 1-7 所示。

表 1-7

航路、进离场航线	英 文 读 法	中 文 读 法
G595	GOLF FIFE NIN-er FIFE	G 五九五 或 GOLF 五九五
A593	ALPHA FIFE NIN-er TREE	A 五九三 或 ALPHA 五九三
VYK-01A	VICTOR YANKEE KILO ZE-RO WUN ALPHA ARRIVAL	大王庄洞幺 ALPHA 进场

练 习 题

1. 口语练习

说明：请按照无线电发音方式读出下列字母组合。

例如：

GUPAD — GOLF YOU NEE FORM PAH PAH AL FAH DELL TAH

SUBUL	DARNA	KAKAT	ANSUK	EPGAM
MAKNO	GOMAN	NONIT	ONEMI	OBLIK

2. 口语练习

说明：请以无线电发音方式读出下列句子。

The quick brown fox jumps over a lazy dog.
Pack my box with five dozen liquor jugs.

3. 听力练习

说明：听写字母与数字组合。

（1）_____.
（2）_____.
（3）_____.
（4）_____.
（5）_____.
（6）_____.
（7）_____.
（8）_____.
（9）_____.
（10）_____.

4. 口语练习

说明：角色扮演，两人一组。首先由驾驶员联系管制员，向其通报航空器呼号、报告点名称、过报告点时间、当前高度、下一报告点名称及预计过下一报告点时间，管制员用"roger"回答。然后交换角色再次练习。例如：

> **P:** Dongfang Control, <u>G-RASC</u>, over <u>GCD</u> at 15, maintaining 9200 meters, estimating <u>SQX</u> at 25.
>
> **C:** <u>G-RASC</u>, roger.

从表 1-8 中选取航空器呼号、报告点名称和下一报告点名称进行替换后练习。

表 1-8

航空器呼号	报告点名称	下一报告点名称
D-ABSP	PLZ	NTX
F-BGJQ	TKG	WOS
I-TMRW	CEH	YIN

1.3 数字

1.3.1 听录音 🔊

表 1-9 为数字的标准读法，请听录音并跟读。

表 1-9

数　字	英 文 读 法	中 文 读 法
0	ZE-RO	洞
1	WUN	幺
2	TOO	两
3	TREE	三
4	FOW-er	四
5	FIFE	五
6	SIX	六
7	SEV-en	拐
8	AIT	八
9	NIN-er	九
.	DAY-SEE-MAL	点
100	HUN-dred	百
1 000	TOU-SAND	千

你能发现有些数字的读法与普通英文里数字的读法不一致吗？发音不同的数字有：

通话中，数字组合可用来表示高度、时间、方向、二次雷达应答机编码、跑道、高度表设定值及频率等，且有其独特的发音规则。数字组合的读法分为一般读法和特殊读法。

1.3.2 数字组合的一般读法

数字组合的英文读法通常按照数字的英文发音及顺序逐位读出；整百、整千或整千整百组合的数字通常读出数字，后面加上"hundred"或"tousand"。

数字组合的中文读法一般按数字的中文发音及顺序逐位读出；整百或整千组合的数字通常读出数字，后面加上"百"或"千"；整千整百组合的数字通常读出数字，千位后面加上"千"，百位后面通常不用加"百"；但一些约定俗成的数字例外，如用于机场区域内描述高度数据的数字"450"，可以读作"四五洞"或"四百五"，如表 1-10 所示。

表 1-10

数 字	英文读法	中文读法
75	SEV-en FIFE	拐五
3 600	TREE TOU-SAND SIX HUN-dred	三千六
45 863	FOW-er FIFE AIT SIX TREE	四五八六三

1.3.3 数字组合的特殊读法

1. 高度

高度分为英制和公制，分别用"英尺"（feet）和"米"（meter）表示。我国高度层配备标准使用公制，在英文通话中需要加上单位"meters"，在中文通话中不加单位"米"，如表 1-11 所示。

表 1-11

高 度 层	英文读法	中文读法
600 m	SIX HUN-dred METERS	六百
900 m	NIN-er HUN-dred METERS	九百
1 200 m	WUN TOU-SAND TOO HUN-dred METERS	幺两
1 500 m	WUN TOU-SAND FIFE HUN-dred METERS	幺五
1 800 m	WUN TOU-SAND AIT HUN-dred METERS	幺八
2 100 m	TOO TOU-SAND WUN HUN-dred METERS	两幺
2 400 m	TOO TOU-SAND FOW-er HUN-dred METERS	两千四
2 700 m	TOO TOU-SAND SEV-en HUN-dred METERS	两拐
3 000 m	TREE TOU-SAND METERS	三千
3 300 m	TREE TOU-SAND TREE HUN-dred METERS	三千三
3 600 m	TREE TOU-SAND SIX HUN-dred METERS	三千六
3 900 m	TREE TOU-SAND NIN-er HUN-dred METERS	三千九

高　度　层	英 文 读 法	中 文 读 法
4 200 m	FOW-er TOU-SAND TOO HUN-dred METERS	四两
4 500 m	FOW-er TOU-SAND FIFE HUN-dred METERS	四千五
4 800 m	FOW-er TOU-SAND AIT HUN-dred METERS	四千八
5 100 m	FIFE TOU-SAND WUN HUN-dred METERS	五幺
5 400 m	FIFE TOU-SAND FOW-er HUN-dred METERS	五千四
5 700 m	FIFE TOU-SAND SEV-en HUN-dred METERS	五拐
6 000 m	SIX TOU-SAND METERS	六千
6 300 m	SIX TOU-SAND TREE HUN-dred METERS	六千三
6 600 m	SIX TOU-SAND SIX HUN-dred METERS	六千六
6 900 m	SIX TOU-SAND NIN-er HUN-dred METERS	六千九
7 200 m	SEV-en TOU-SAND TOO HUN-dred METERS	拐两
7 500 m	SEV-en TOU-SAND FIFE HUN-dred METERS	拐五
7 800 m	SEV-en TOU-SAND AIT HUN-dred METERS	拐八
8 100 m	AIT TOU-SAND WUN HUN-dred METERS	八幺
8 400 m	AIT TOU-SAND FOW-er HUN-dred METERS	八千四
8 900 m	AIT TOU-SAND NIN-er HUN-dred METERS	八千九
9 200 m	NIN-er TOU-SAND TOO HUN-dred METERS	九千二
9 500 m	NIN-er TOU-SAND FIFE HUN-dred METERS	九千五
9 800 m	NIN-er TOU-SAND AIT HUN-dred METERS	九千八
10 100 m	WUN ZE-RO TOU-SAND WUN HUN-dred METERS	幺洞幺
10 400 m	WUN ZE-RO TOU-SAND FOW-er HUN-dred METERS	幺洞四
10 700 m	WUN ZE-RO TOU-SAND SEV-en HUN-dred METERS	幺洞拐
11 000 m	WUN WUN TOU-SAND METERS	幺幺洞
11 300 m	WUN WUN TOU-SAND TREE HUN-dred METERS	幺幺三
11 600 m	WUN WUN TOU-SAND SIX HUN-dred METERS	幺幺六
11 900 m	WUN WUN TOU-SAND NIN-er HUN-dred METERS	幺幺九
12 200 m	WUN TOO TOU-SAND TOO HUN-dred METERS	幺两两
12 500 m	WUN TOO TOU-SAND FIFE HUN-dred METERS	幺两五
13 100 m	WUN TREE TOU-SAND WUN HUN-dred METERS	幺三幺
13 700 m	WUN TREE TOU-SAND SEV-en HUN-dred METERS	幺三拐
14 300 m	WUN FOW-er TOU-SAND TREE HUN-dred METERS	幺四三
14 900 m	WUN FOW-er TOU-SAND NIN-er HUN-dred METERS	幺四九

　　以标准大气压 1 013.2 hPa 为基准面，对符合英制高度层配备标准的高度，使用英文读法时，按照国际民航组织规定的发音，在 "flight level" 后逐位读出万位、千位和百位上的数字；

使用中文读法时，读出万位、千位和百位上的数字，高度层低于 10 000 英尺时，读作"×千英尺"，如表 1-12 所示。

表 1-12

高 度 层	英 文 读 法	中 文 读 法
9 000 ft	FLIGHT LEVEL NIN-er ZE-RO	九千英尺
33 000 ft	FLIGHT LEVEL TREE TREE ZE-RO	三三洞

当高度指令涉及气压基准面转换时，管制员应在通话中指明新的气压基准面数值，在以后的通话中可省略气压基准面。使用英文读法时，对上升到 1 013.2 hPa 为基准面的高度，在高度数字后加上"on standard"，当以修正海平面气压为基准面时，在高度数字后加上"on QNH ××××（数值）"；当以场面气压为基准面时，在高度数字后加上"on QFE ××××（数值）"。

使用中文读法时，对上升到以 1 013.2 hPa 为基准面的高度，在高度数字前加上"标准气压"；当以修正海平面气压为基准面时，在高度数字前加上"修正海压"，在高度数字后加上修正海压数值；当以场面气压为基准面时，在高度数字前加上"场压"，在高度数字后加上场压数值。例如：

> **C:** HDA905, descend to 1800 meters on QNH 1011.
>
> **C:** HDA905，下降到修正海压幺八，修正海压幺洞幺幺。

2. 时间

时间的读法一般只读出分，必要时（如可能引起混淆的情况下）应读出小时和分，如表 1-13 所示。

表 1-13

时　　间	英 文 读 法	中 文 读 法
0803（上午 8:03）	ZE-RO TREE　or　ZE-RO AIT ZE-RO TREE	洞三或洞八洞三
1300（下午 1:00）	ZE-RO ZE-RO　or　WUN TREE ZE-RO ZE-RO	整点或幺三洞洞

当驾驶员觉得必要时可向管制员申请校对时间（request time check），此时将实际时间半分钟向上取整，通报时间的方式如表 1-14 所示。

表 1-14

时　　间	英 文 读 法	中 文 读 法
11:55:18	WUN WUN FIFE FIFE AND A HALF	幺幺五五三洞
11:55:38	WUN WUN FIFE SIX	幺幺五六

3. 气压

数字应逐位读出，如表 1-15 所示。

表 1-15

气　压	英 文 读 法	中 文 读 法
QFE 1003	QFE WUN ZE-RO ZE-RO TREE	场压幺洞洞三
QNH 1000	QNH WUN ZE-RO ZE-RO ZE-RO①	修正海压幺洞洞洞

4. 航向

航向后应跟三位数并逐位读出数值，如表 1-16 所示。

表 1-16

航　向	英 文 读 法	中 文 读 法
100°	HEADING WUN ZE-RO ZE-RO	航向幺洞洞
005°	HEADING ZE-RO ZE-RO FIFE	航向洞洞五
360°	HEADING TREE SIX ZE-RO	航向三六洞

5. 速度

英文读法：逐位读出数值，后加速度单位；马赫数的读法为省略小数点前的 0，读作"mach point ××"。

中文读法：使用海里每小时（节）作为速度单位时，逐位读出数值，后不加单位；使用马赫数作为速度单位时，读作"马赫数点××"或"马赫数×点××"，如表 1-17 所示。

表 1-17

速　度	英 文 读 法	中 文 读 法
280 knots	TOO AIT ZE-RO KNOTS	两八洞
M0.85	MACH POINT AIT FIFE	马赫数点八五

6. 风向风速

风向风速通常作为一个整体通报，英文读法中风向后需加"degrees"，风速单位为 knots，中文读法中一般使用米/秒（m/s），如表 1-18 所示。

表 1-18

风 向 风 速	英 文 读 法	中 文 读 法
160°, 7 m/s, gusting 15 m/s	[SURFACE] WIND WUN SIX ZE-RO DEGREES SEV-en METERS PER SECOND GUSTING WUN FIFE METERS PER SECOND	地面风幺六洞，拐米秒，阵风幺五米秒

---

① ICAO Annex10 中高度表设定值的英文读法更新为：高度表设定值应逐位读出，除了 1 000 hPa。1 000 hPa 的读法应为 ONE TOUSAND。但目前我国一线运行中 1 000 hPa 的英文读法还是逐位读出。

7. 频率

频率应逐位读出，且英文读法应读出高频的单位，如表 1-19 所示。

表 1-19

频　率	英 文 读 法	中 文 读 法
121.45 MHz	WUN TOO WUN DAY-SEE-MAL FOW-er FIFE	幺两幺点四五
6 565 kHz	SIX FIFE SIX FIFE KILO HERTZ	六五六五

根据甚高频通信波道拥挤的状况，国际民航组织在某些地区将甚高频通信波道的频率间隔由 25 kHz 减小为 8.33 kHz。使用 8.33 kHz 频率间隔的频率由六位数字组成，当最后两位均为"0"时，只需读出前四位，否则应逐位读出所有六位数字，如表 1-20 所示。

表 1-20

波　道	英 文 读 法	中 文 读 法
118.000 MHz	WUN WUN AIT DAY-SEE-MAL ZE-RO	幺幺八点洞
118.010 MHz	WUN WUN AIT DAY-SEE-MAL ZE-RO WUN ZE-RO	幺幺八点洞幺洞
118.025 MHz	WUN WUN AIT DAY-SEE-MAL ZE-RO TOO FIFE	幺幺八点洞两五

8. 跑道

跑道编号应按照数字的英文或中文发音逐位读出。跑道编号后的英文字母"R""L""C"，英文读作"right""left""center"，中文读作"右""左""中"，如表 1-21 所示。

表 1-21

跑　道　号	英 文 读 法	中 文 读 法
03	RUNWAY ZE-RO TREE	跑道洞三
08L	RUNWAY ZE-RO AIT LEFT	跑道洞八左

9. 距离

按照数字的一般读法读出，后面加上单位，如表 1-22 所示。

表 1-22

距　离	英 文 读 法	中 文 读 法
18 n mile	WUN AIT MILES[①]	幺八海里
486 km	FOW-er AIT SIX KILOMETERS	四八六公里
400 km	FOW-er HUN-dred KILOMETERS	四百公里

10. 应答机编码

应答机编码按照数字的一般读法逐位读出，如表 1-23 所示。

---

① 在陆空通话中，表示距离的单位是海里（n mile），英文读法约定俗成读作"mile"。

表 1-23

应答机编码	英 文 读 法	中 文 读 法
3213	SQUAWK　TREE TOO WUN TREE①	应答机三两幺三
5731	SQUAWK　FIFE SEV-en TREE WUN	应答机五拐三幺

11. 航空器机型

航空器机型通常按照航空器制造商注册名加机型的方式读出，如表 1-24 所示。

表 1-24

机　　型	英 文 读 法	中 文 读 法
B737-800/B738	BOEING SEV-en TREE SEV-en AIT HUN-dred/Boeing SEV-en TREE AIT	波音七三七八百/波音七三八
A340	AIRBUS TREE FOW-er ZE-RO	空客三四零
MA60	MODERN ARK SIX ZE-RO	新舟六零
EMB190	EMB WUN NIN-er ZE-RO	EMB 幺九零
TU-204	TUPOLEV TOO ZE-RO FOW-er	图两洞四
D-328	DORNIER TREE TOO AIT	道尼尔三两八
CRJ-200	CRJ TOO HUN-dred	CRJ 两百

12. 气象

有关气象方面的数字的读法，如能见度和跑道视程，按照表 1-25 所示读出。

表 1-25

气　　象	英 文 读 法	中 文 读 法
能见度 2 000	VISIBILITY TOO TOU-SAND METERS	能见度两千米
跑道视程 700	RUNWAY VISUAL RANGE (or RVR) SEV-en HUN-dred METERS	跑道视程七百米

13. 经纬度

经纬度的英文读法为："× DEGREES × MINUTES × SECONDS" + "NORTH (SOUTH)"；"× DEGREES × MINUTES × SECONDS" + "EAST (WEST)"。

经纬度的中文读法为："北（南）纬" + "×度×分×秒"；"东（西）经" + "×度×分×秒"，如表 1-26 所示。

表 1-26

经　纬　度	英 文 读 法	中 文 读 法
62° 08′ 12″ N, 137° 32′ 06″ E	Coordinates SIX TOO DEGREES ZE-RO AIT MINUTES WUN TOO SECONDS NORTH WUN TREE SEN-en DEGREES TREE TOO MINUTES ZE-RO SIX SECONDS EAST	北纬六两洞八幺两，东经幺三拐三两洞六
25° 21′ 15″ S, 147° 15′ 25″ W	Coordinates TOO FIFE DEGREES TOO WUN MINUTES WUN FIFE SECONDS SOUTH WUN FOW-er SEN-en DEGREES WUN FIFE MINUTES TOO FIFE SECONDS WEST	南纬两五两幺幺五，西经幺四拐幺五两五

① ICAO Annex10 中应答机编码的英文读法更新为：编码应逐位读出，除了编码中包含整千的信息。包含整千的编码的读法为："数字" + "TOUSAND"。例如，2000 的英文读法为 "squawk two thousand"。

练 习 题

1. 口语练习

说明：请按照英文读法读出下列数字。

13257	23089	49000	56782	11004
32200	78936	89001	97538	66450

2. 口语练习

说明：根据图 1-1 中飞行高度层配备标准，按照标准读法读出米制高度层。

飞行高度层配备标准示意图

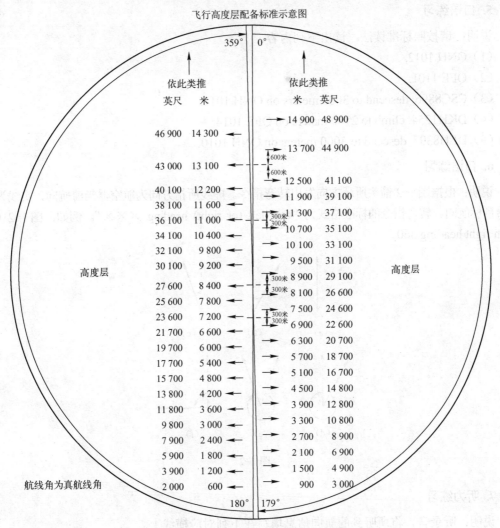

图 1-1　飞行高度层配备标准示意图

3. 听力练习

说明：听录音，将所听到的高度和气压基准面信息填写在下列对应横线上。

（1）_____.
（2）_____.
（3）_____.
（4）_____.
（5）_____.

4. 口语练习

说明：按照数字的标准读法，读出下列时间。

| 0700 | 0945 | 1000 | 1330 | 1415 |
| 0001 | 0030 | 2020 | 0856 | 1111 |

5. 口语练习

说明：请按照标准读法，读出下列内容。

（1）QNH 1012.

（2）QFE 1101.

（3）CSC8896, descend to 3000 meters on QNH 1011.

（4）DKH1254, climb to 2700 meters on QFE 1014.

（5）DLH8397, descend to 3000 meters on QNH 1010.

6. 口语练习

说明：根据图 1-2 箭头所示方向发出转弯指令。虚线所指方向为航空器当前航向，实线为航空器目标航向。转弯指令的标准格式为 "turn left (or right) heading ×××"。例如，图 1-2（a）"turn right heading 360"。

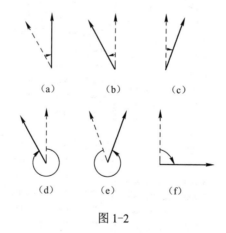

图 1-2

7. 听力练习

说明：听录音，将所听到的航向信息填写在下列对应横线上。

（1）_____.
（2）_____.
（3）_____.

（4）_____.

（5）_____.

8. 口语练习

说明：请按照速度的标准读法，给出调速指令。调速指令的标准格式为"reduce (or increase) speed to ×××"。

（1）reduce speed to 250 knots.

（2）reduce speed to 180 knots.

（3）reduce speed to M0.75.

（4）reduce speed to M0.85.

（5）increase speed to 330 knots.

9. 口语练习

说明：请按照标准读法，发出下列指令。

（1）CES5183, Pudong Ground, departure Runway 35R, wind 250 degrees, 10 m/s, QNH 1015, temperature 25, dew point 23, RVR 500 meters.

（2）AFR129, wind 120 degrees, 10 knots, gusting to 15 knots, Runway 23, cleared for take-off.

（3）CSH9134, wind 240 degrees, 12 m/s.

（4）CPA347, Runway 15, wind variable, 5 knots, cleared to land.

（5）CQN2302, wind 330 degrees, 10 m/s.

10. 听力练习

说明：听录音，将所听到的频率信息填写在下列对应横线上。

（1）_____.

（2）_____.

（3）_____.

（4）_____.

（5）_____.

11. 口语练习

说明：按照标准读法，读出下列指令。

（1）KAL830, wind 160 degrees, 4 m/s, Runway 23, cleared to land.

（2）ANA956, wind 210 degrees, 5 m/s, Runway 36L, cleared for take-off.

（3）DAL128, wind 200 degrees, 10 m/s, Runway 18R, cleared to land.

（4）CSZ9403, wind 230 degrees, 7 m/s, Runway 27, cleared for take-off.

（5）CBJ5357, wind 120 degrees, 9 m/s, Runway 19, cleared to land.

12. 口语练习

说明：按照标准读法，读出下列内容。

（1）20 kilometers.

（2）8 miles.

（3）CES2105, 10 kilometers from touchdown, radar service terminated, contact tower 118.1, good day.

（4）KLM898, position 12 miles west of VYK.

（5）UAL889, position 10 miles east of the field.

13. 听力练习

说明：听录音，将所听到的应答机编码信息填写在下列对应横线上。

（1）＿＿＿＿＿＿＿＿＿＿＿＿＿＿＿＿＿＿＿＿＿＿＿＿＿＿＿＿＿＿＿＿.

（2）＿＿＿＿＿＿＿＿＿＿＿＿＿＿＿＿＿＿＿＿＿＿＿＿＿＿＿＿＿＿＿＿.

（3）＿＿＿＿＿＿＿＿＿＿＿＿＿＿＿＿＿＿＿＿＿＿＿＿＿＿＿＿＿＿＿＿.

（4）＿＿＿＿＿＿＿＿＿＿＿＿＿＿＿＿＿＿＿＿＿＿＿＿＿＿＿＿＿＿＿＿.

（5）＿＿＿＿＿＿＿＿＿＿＿＿＿＿＿＿＿＿＿＿＿＿＿＿＿＿＿＿＿＿＿＿.

14. 听力练习

说明：听录音，将所听到的航空器机型信息填写在下列对应横线上。

（1）＿＿＿＿＿＿＿＿＿＿＿＿＿＿＿＿＿＿＿＿＿＿＿＿＿＿＿＿＿＿＿＿.

（2）＿＿＿＿＿＿＿＿＿＿＿＿＿＿＿＿＿＿＿＿＿＿＿＿＿＿＿＿＿＿＿＿.

（3）＿＿＿＿＿＿＿＿＿＿＿＿＿＿＿＿＿＿＿＿＿＿＿＿＿＿＿＿＿＿＿＿.

（4）＿＿＿＿＿＿＿＿＿＿＿＿＿＿＿＿＿＿＿＿＿＿＿＿＿＿＿＿＿＿＿＿.

（5）＿＿＿＿＿＿＿＿＿＿＿＿＿＿＿＿＿＿＿＿＿＿＿＿＿＿＿＿＿＿＿＿.

15. 听力练习

说明：听录音，将所听到的气象信息填写在下列对应横线上。

（1）＿＿＿＿＿＿＿＿＿＿＿＿＿＿＿＿＿＿＿＿＿＿＿＿＿＿＿＿＿＿＿＿.

（2）＿＿＿＿＿＿＿＿＿＿＿＿＿＿＿＿＿＿＿＿＿＿＿＿＿＿＿＿＿＿＿＿.

（3）＿＿＿＿＿＿＿＿＿＿＿＿＿＿＿＿＿＿＿＿＿＿＿＿＿＿＿＿＿＿＿＿.

（4）＿＿＿＿＿＿＿＿＿＿＿＿＿＿＿＿＿＿＿＿＿＿＿＿＿＿＿＿＿＿＿＿.

（5）＿＿＿＿＿＿＿＿＿＿＿＿＿＿＿＿＿＿＿＿＿＿＿＿＿＿＿＿＿＿＿＿.

1.4　标准单词和词组

下列标准单词在通话中具有特定的含义。

（1）acknowledge（请认收）——Let me know that you have received and understood this message. （向我表示你已经收到并理解该电报。）

例如：

> **C:** CHB6263, do not acknowledge my further transmissions.
>
> **C:** CHB6263，请不要认收以下通话。

（2）affirm（是的）——Yes. （是的。）

例如：

C: CSN6412, are you ready for immediate departure?

P: Affirm, CSN6412.

C: CSN6412，能立即起飞吗？

P: 是的，CSN6412。

（3）approved（同意）——Permission for proposed action granted.（批准所申请的行动。）

例如：

P: Dongfang Ground, UAL7610, request push-back and start up.

C: UAL7610, push-back and start up approved.

P: 东方地面，UAL7610，请求推出开车。

C: UAL7610，同意推出开车。

（4）break（还有）——I hereby indicate the separation between portions of the message to be used where there is no clear distinction between the text and other portions of the message.（表示电报各部分的间断；用于电文与电报的其他部分无明显区别的情况。如果信息的各个部分之间没有明显的区别可以使用该词作为信息各部分之间的间隔标志。）

例如：

C: ACA030, taxi to holding point A1, Runway 18R, break, taxiway centreline lighting unserviceable.

C: ACA030，滑到 A1 等待点，跑道 18 右，还有，滑行道中央灯光故障。

（5）break break（另外）——I hereby indicate the separation between messages transmitted to different aircraft in a very busy environment. （表示在非常繁忙的情况下，发布给不同航空器的电报之间的间断。）

例如：

C: AAL186, descend immediately to 6900 meters, break break, CCA1421, climb immediately to 7500 meters.

C: AAL186，立即下降到六千九，另外，CCA1421，立即上升到拐五。

（6）cancel（取消）——Annul the previously transmitted clearance.（废除此前所发布的许可。）

例如：

> **C:** THY021, hold position, cancel take-off, I say again, cancel take-off, vehicle crossing the runway.
>
> **C:** THY021，原地等待，取消起飞，重复一遍，取消起飞，有车辆穿越跑道。

（7）check（检查）——Examine a system or procedure, and no answer is normally expected.（检查系统或程序，且通常不回答。）

例如：

> **C:** KAL5802, check your transmitter and give me a long call.
>
> **C:** KAL5802，检查你的发射机并给我一长呼。

（8）cleared（可以）——Authorized to proceed under the conditions specified.（批准按指定条件前行。）

例如：

> **C:** JAL020, wind calm, Runway 18, cleared for take-off.
>
> **C:** JAL020，静风，跑道 18，可以起飞。

（9）confirm（证实）——I request verification of (clearance, instruction, action, information)（我请求核实：许可、指令、行动和信息。）

例如：

> **C:** SWR197, confirm squawk 3352.
>
> **C:** SWR197，证实应答机编码 3352。

（10）contact（联系）——Establish communications with...（与……建立通信。）

例如：

> **C:** SIA805, contact Dongfang Tower on 118.8.
>
> **C:** SIA805，联系东方塔台 118.8。

（11）correct（正确）——True or accurate.（真实的或准确的。）

例如：

> **C:** UAE307, read-back correct.
>
> **C:** UAE307，复诵正确。

（12）correction（更正）——An error has been made in this transmission (or message indicated). The correct version is...（本电报出了一个错误，或所发布的信息本身是错的，正确的内容应当是……）

例如：

> **C:** THA615, climb to 7500 meters, correction, climb to 7800 meters.
>
> **C:** THA615，上升到拐五，更正，上升到拐八。

（13）disregard（作废）——Ignore.（忽视。）

例如：

> **C:** MAS361, descend to…, disregard, maintain present level.
>
> **C:** MAS361，下降到……，作废，保持现在高度。

（14）How do you read?（你听我几个？）——What is the readability of my transmission?（我所发电报的清晰度如何？）

例如：

> **P:** Dongfang Tower, MGL224, how do you read?
>
> **P:** 东方塔台，MGL224，你听我几个？

（15）I say again（我重复一遍）——I repeat for clarity or emphasis.（为了表示澄清或强调，我重复一遍。）

例如：

> **C:** PAL359, hold position, cancel take-off, I say again, cancel take-off, due vehicle on the runway.
>
> **C:** PAL359，原地等待，取消起飞，重复一遍，取消起飞，跑道上有车辆。

（16）maintain（保持）——Continue in accordance with the condition(s) specified or in its literal sense, e.g. "maintain VFR".（依照指定条件继续或按字面意义，如"保持 VFR"。）

例如：

> **C:** CHH7265, maintain 5400 meters, expect descent after VYK.
> **P:** Maintaining 5400 meters, CHH7265.
>
> **C:** CHH7265，保持五千四，预计过大王庄后下降。
> **P:** 保持五千四，CHH7265。

（17）monitor（守听）——Listen out on (frequency).（收听或调定到某个频率。）

例如：

> **C:** PIA853, monitor ATIS 123.25.
>
> **C:** PIA853，123.25 上守听通播。

（18）negative（错误、不同意或没有）——No or Permission not granted or That is not correct or Not capable.（并非如此，或不允许，或不对，或不能。）

例如：

> **C:** QTR895, descend to 8400 meters.
> **P:** Descending to 8100 meters, QTR895.
> **C:** QTR895, negative, descend to 8400 meters.
> **P:** Descending to 8400 meters, QTR895.
>
> **C:** QTR895，下降到八千四。
> **P:** 下降到八千一，QTR895。
> **C:** QTR895，不对，下降到八千四。
> **P:** 下降到八千四，QTR895。

（19）out（完毕）——This exchange of transmissions is ended and no response is expected.（本次通话已经结束，并且你不需要作出回答。）

"out" 通常不用于 VHF 通信中。

（20）over（请回答）——My transmission is ended and I expect a response from you.（我发话完毕，并希望你回答。）

"over" 通常不用于 VHF 通信中。

（21）read back（复诵）——Repeat all, or the specified part of this message back to me exactly as received.（请向我准确地重复本电报所有或部分内容。）

例如：

> **C:** AMU007, descend to 2700 meters on QNH 1010.
> **P:** Descending to 2700 meters, AMU007.
> **C:** AMU007, please read back QNH.
>
> **C:** AMU007，下降到修正海压两拐，修正海压幺洞幺洞。
> **P:** 下降到修正海压两拐，AMU007。
> **C:** AMU007，复诵修正海压。

（22）recleared（重新许可）——A change has been made to your last clearance and this new clearance supersedes your previous clearance or part thereof.（此前发布给你的许可已经变更，这一新的许可将取代刚才的许可或其中部分内容。）

例如：

> **C:** OKA2851, recleared to Dongfang via KNOCK J1 BRAVO.

> **C:** OKA2851，重新许可经由 KNOCK J1 BRAVO 去往东方机场。

（23）report（报告）——Pass me the following information...（向我报告下列信息……）

例如：

> **C:** HXA2674, report speed.
> **P:** Speed 250 knots, HXA2674.
>
> **C:** HXA2674，报告速度。
> **P:** 速度 250，HXA2674。

（24）request（请求）——I should like to know... or I wish to obtain...（我希望知道……或我希望得到……）

例如：

> **P:** Dongfang Ground, HVN513, request taxi to Runway 36L.
>
> **P:** 东方地面，HVN513，请求滑行到 36 左跑道。

（25）roger（收到）——I have received all of your last transmission.（我已经收到了你刚才的发话。）

在任何情况下，不得采用"roger"来回答要求复诵或要求回答"是"或"否"的问题。

例如：

> **C:** CAL512, a preceding aircraft reported wind shear on final.
> **P:** Roger, CAL512.
>
> **C:** CAL512，前机报告五边有风切变。
> **P:** 收到，CAL512。

（26）say again（重复一遍）——Repeat all, or the following part of your last transmission.（请重复你刚才发话的所有内容或下列部分。）

例如：

> **P:** Dongfang Tower, ...
> **C:** Station calling Tower, say again your call sign.
> **P:** Dongfang Tower, AFL201.
>
> **P:** 东方塔台，……
> **C:** 哪个呼叫塔台，重复一遍你的呼号。
> **P:** 东方塔台，AFL201。

（27）speak slower（讲慢点）——Reduce your rate of speech.（请降低你的语速。）

例如：

> **C:** FIN052, Dongfang Tower, speak slower.
>
> **C:** FIN052，东方塔台，请讲慢点。

（28）stand by（稍等）——Wait and I will call you.（请等候，我将呼叫你。）
例如：

> **P:** Dongfang Tower, AAR336, Gate 15, Information C, request push-back.
> **C:** AAR336, stand by.
> **P:** Standing by, AAR336.
>
> **P:** 东方塔台，AAR336，停机位 15，通播 C，请求推出。
> **C:** AAR336，稍等。
> **P:** 稍等，AAR336。

（29）unable 或 unable to comply（无法执行）——I cannot comply with your request, instruction, or clearance.（我不能按照你的请求、指令或许可执行。）

unable 后通常应跟不能执行的原因。
例如：

> **C:** CES5186, climb to 8900 meters, expedite until passing 8400 meters.
> **P:** Unable to expedite climb due weight, CES5186.
>
> **C:** CES5186，上升到八千九，尽快通过八千四。
> **P:** 由于重量大，不能尽快通过八千四，CES5186。

（30）wilco（照办）——Abbreviation for "will comply". I understand your message and will comply with it.（"将照办"的缩略语。我已经明白了你的电报并将按照该电报执行。）
例如：

> **C:** EVA716, report over ZHO.
> **P:** Wilco, EVA716.
>
> **C:** EVA716，到达周口报告。
> **P:** 照办，EVA716。

（31）words twice（讲两遍）—— ① As a request: Communication is difficult. Please send every word or group of words twice. [对于申请来说：因为通信困难，请把每个词（组）发送两遍。] ② As information: Since communication is difficult, every word or group of words in this message will be sent twice. [对于信息来说：因为通信困难，该电报的每个词（组）将被发送两遍。]

例如：

> C: ALK504, here is your ATC clearance, words twice.
> P: Ready to copy, ALK504.
>
> C: ALK504，你的放行许可，讲两遍。
> P: 准备抄收，ALK504。

练 习 题

1. 词汇练习

说明：将左侧标准单词与右侧中文含义连线配对。

acknowledge	可以
correction	请认收
disregard	是的
affirm	同意
correct	证实
confirm	联系
approved	正确
cleared	更正
monitor	守听
contact	作废

2. 完成通话

说明：用恰当的单词或词组填空完成通话。

（1）Do not _____ my further messages.

（2）Push-back and start up _____.

（3）HDA903, Runway 36L, _____ to land.

（4）CCA1346, descend to 900 meters, _____, Taxiway X closed.

（5）CSN3137, turn right heading 090 immediately, _____, CES3074, turn right heading 270 immediately.

（6）C: AFR381, radar service terminated, 14 kilometers from touchdown, _____ Beijing Tower 119.1, good day.

    P: 119.1, good day, AFR381.

（7）C: SAS995, _____ QNH 1012.

    P: Affirm, QNH 1012, SAS995.

（8）C: CCA1331, climb to 4800 meters on QNH, _____, climb to 4800 meters on standard.

    P: Climbing to 4800 meters on QNH, CCA1331.

C: ＿＿＿＿, climb to 4800 meters on standard.

P: Climbing to 4800 meters on standard, CCA1331.

C: CCA1331, ＿＿＿＿＿＿＿.

（9）Shanghai approach, CPA5858, 2700 meters maintaining, ＿＿＿＿＿＿＿＿＿＿＿＿＿, 2700 meters maintaining, engine losing power, left engine losing power.

（10）**C:** KLM4302, are you ready for departure?

　　　　P: ＿＿＿＿＿＿, KLM4302.

3. 词汇练习

说明：举例辨析下列标准单词或词组。

affirm 与 correct	approved 与 cleared
confirm 与 check	say again 与 I say again
break 与 break break	correct 与 correction
roger 与 acknowledge	stand by 与 monitor

4. 听力练习

说明：听录音，按照表 1-27 中（1）所示，把其余空白处补充完整。

表 1-27

驾驶员以为听到的是	管制员实际说的是
（1）climb to 22500 feet	climb to 2500 feet
（2）	
（3）	
（4）	
（5）	

5. 思考题

（1）陆空通话中可能造成误解的原因有哪些？

（2）写出陆空通话中（英文或中文）发音相似的数字、单词或词组。

1.5　呼号

1.5.1　管制单位呼号

管制单位一般用地名加后缀的方式作为其呼号，后缀表明提供何种服务或单位类型，如表 1-28 所示。

表 1-28

管制或服务单位	呼 号 后 缀
area control center（区域管制中心）	CONTROL（区域）
radar（雷达①）	RADAR（雷达）
approach control（进近管制）	APPROACH（进近）
approach control radar arrival（进场雷达管制）	ARRIVAL（进场）
approach control radar departure（离场雷达管制）	DEPARTURE（离场）
aerodrome control（机场管制）	TOWER（塔台）
surface movement control（地面活动管制）	GROUND（地面）
clearance delivery（放行许可发布）	DELIVERY（放行）
precision approach radar（精密进近雷达）	PRECISION（精密）
flight information service（飞行情报服务）	INFORMATION（飞服）
apron control/management service（机坪管制或管理服务）	APRON（机坪）
company dispatch（公司签派）	DISPATCH（签派）
aeronautical station（航空电台）	RADIO（电台）
final approach（五边）	FINAL（五边）

① 此为通用雷达。

练 习 题

说明：请从下框中选取适当内容完成通话。

> BAW038　　CES2105　　CCA1108　　DKH1254　　CXA8130　　Dongfang Tower
> Dongfang Approach　　Dongfang Delivery　　Dongfang Departure　　Dongfang Control
> climb to 4800 meters　　descend to 9500 meters　　request taxi
> request ATC clearance　　cleared for ILS approach

（1）_____.
（2）_____.
（3）_____.
（4）_____.
（5）_____.

1.5.2　航空器呼号

航空器呼号分为三类，其中某些形式的呼号有简呼形式。

（1）航空器的注册号：注册号字母和数字应按照字母和数字的标准发音逐位读出。有时以航空器制造厂商或航空器机型名称作为注册号字母的前缀，制造厂商或航空器机型名称则按照英文发音习惯或翻译的中文读出。例如：

> G-ABCD: GOLF ALPHA BRAVO CHARLIE DELTA
>
> Cessna G-ABCD: CESSNA GOLF ALPHA BRAVO CHARLIE DELTA

（2）航空器经营人的无线电呼号加航空器注册号的最后四位字母：航空器经营人呼号的英文发音按照 ICAO 指定的无线电呼号读出，中国航空公司呼号的中文发音按照中国民航规定的呼号读出；注册号的字母全部按照字母的英文标准发音逐位读出；数字应分别按照数字的英文、中文标准发音逐位读出。例如：

> BAW BHWC: SPEEDBIRD BRAVO HOTEL WHISKEY CHARLIE

（3）航空器经营人的无线电呼号加航班号：航空器经营人呼号的英文发音按照 ICAO 指定的无线电呼号读出，中国航空公司呼号中文发音按照中国民航规定的呼号读出；航班号的字母全部按照字母的英文标准发音逐位读出；数字应按照数字的英文、中文标准发音逐位读出。例如：

> BAW038: SPEEDBIRD ZE-RO TREE AIT
>
> CES7255: CHINA EASTERN SEV-en TOO FIFE FIFE（英文读法）　　or　　东方拐两五五（中文读法）

在建立满意的双向通信联系之后，在无任何混淆产生的情况下，可以使用航空器的简呼，且先应由管制员简呼航空器，航空器才能以简呼作为自呼。上述航空器呼号的缩减形式如下。

（1）航空器注册号中第一个字符和至少最后两个字符（字母或数字），如：N 57826 简呼为 N26 或 N826；G-ABCD 简呼为 G CD 或 G BCD；CESSNA FABCD 简呼为 CESSNA CD 或 CESSNA BCD。可以用航空器制造厂商或航空器机型名称替换航空器注册号中第一个字符。

（2）航空器经营人的无线电呼号加航空器注册号中至少最后两个字符，如：CCA BHWC 简呼为 CCA WC 或 CCA HWC。

（3）航空器经营人的无线电呼号加航班号，无缩减形式，如 BAW038 无简呼。

当由于存在相似呼号而可能产生混淆时，管制单位可临时指示航空器改变呼号形式。例如：

> **C:** CSN3107, change your call sign to B2456, similar call sign.
>
> **P:** Change call sign to B2456, CSN3107.
>
> 　(A moment later)
>
> **C:** B2456, revert to flight plan call sign CSN3107 at NSH.
>
> **C:** CSN3107，将你的呼号改为 B2456，有相似呼号。
>
> **P:** 将呼号改为 B2456，CSN3107。
>
> 　(稍后)
>
> **C:** B2456，在宁陕恢复到飞行计划中的呼号 CSN3107。

　　对于 A380 机型，当机组与管制单位首次建立联系时，驾驶员必须在其航班呼号后增加"super"一词。当航空器尾流等级为重型时，在与塔台和进近管制员首次联系时应在其呼号后加上"重型"（heavy）一词。例如：

> **P:** Dongfang Tower, UAL089, heavy, request taxi.
>
> **P:** 东方塔台，UAL089，重型，请求滑行。

练 习 题

1. 口语练习

说明：角色扮演，两人一组。航空器离地后，首先由驾驶员联系管制员，管制员向其发出高度指令；然后交换角色再次练习。例如：

> **P:** Beijing Approach, G-ABDC, airborne 15,150 meters climbing.
>
> **C:** G-ABDC, Beijing Approach, climb to and maintain 1200 meters.
>
> **P:** Climb to and maintain 1200 meters, G-ABDC.
> 　　(A moment later)
>
> **C:** G-ABDC, continue climb to 4800 meters on standard.
>
> **P:** Continue climb to 4800 meters on standard, G-ABDC.

从表 1-29 中选取管制单位呼号和航空器呼号进行替换后练习。

表 1-29

管制单位呼号	航空器呼号
Dongfang Approach	F-FTUM
Guangzhou Approach	G-TSPL
Tianjin Approach	I-CGHQ
Shanghai Approach	D-ABSP
Wuhan Approach	F-BGJQ

2. 完成通话

说明：听录音，完成通话。

（1）**P:** Beijing Approach, _____, Information _____, _____maintaining.

　　　C: _____, maintain _____. Report passing HUR.

　　　P: Maintain 5400 meters. Will report passing HUR. _____.

（2）**P:** Guangzhou Approach, _____.

C: ＿＿＿＿＿＿＿, Guangzhou Approach, report heading and level.

P: Heading ＿＿＿＿＿＿, ＿＿＿＿＿＿, ＿＿＿＿＿＿.

（3）C: ＿＿＿＿＿＿＿＿＿＿＿＿＿＿＿＿＿＿＿＿＿＿＿＿＿.

P1: Descend to and maintain 2400 meters on QNH 1004, CCA1357.

C: ＿＿＿＿＿＿＿＿＿＿＿＿＿＿＿＿＿＿＿＿＿＿＿＿＿.

P1: Maintain present level, CCA1357.

C: ＿＿＿＿＿＿＿＿＿＿＿＿＿＿＿＿＿＿＿＿＿＿＿＿＿.

P2: Descend to and maintain 2400 meters on QNH 1004, CCA1375.

C: ＿＿＿＿＿＿＿＿＿＿＿＿＿＿＿＿＿＿＿＿＿＿＿＿＿.

3. 听力练习

说明：请先读出下列航空器呼号，然后听录音，在所听到的航空器呼号下画横线。

CHH7761	QTR803	JAL883	CDG4676
HVN6416	CES5263	UAL858	HDA811
SIA831	CDG4646	KAL893	CSN3554
JAL891	CQH8876	CES2668	CSH9206
QTR807	UAL835	CSC8954	CES5623
CSN3524	CSH9260	JAL81	HVN416
UAE309	CSC8974	CQH8867	CHH7721

1.6　通信的基本方法

1.6.1　通信的建立

初次建立联系时，航空器应使用航空器和管制单位的全称。例如：

P: Shanghai Approach, G-ABCD.

C: G-ABCD, Shanghai Approach.

P: 上海进近，G-ABCD。

C: G-ABCD，上海进近。

如果地面电台或某一航空器需要广播信息或情报时，可以在信息或情报前加上"All stations"（全体注意）。例如：

P: All stations, G-CDAB, southbound CHO VOR to Dongfang, leaving 6000 meters now, descending to 4500 meters.

P: 全体注意，G-CDAB，从 CHO VOR 往南去东方，现在离开六千，下降到四千五。

在不确定信息是否接收正确时，可要求重复所发送信息的部分或全部内容，采用表1-30

中的用语。

表 1-30

用　语	含　义
say again（重复）	repeat entire message（重复完整信息）
say again…（item）（重复……）	repeat specific item（重复特定信息项）
say again all before…（重复……之前）	repeat that part of the message before the first satisfactorily received word（重复第一个接收清楚单词之前的那部分信息）
say again all after…（重复……之后）	repeat that part of the message after the last satisfactorily received word（重复最后一个接收清楚单词之后的那部分信息）
say again all between… and…（重复……和……之间）	repeat that part of the message between two satisfactorily received words（重复在两个接收清楚单词之间的那部分信息）

例如：

C: VIR7937, cleared to Hong Kong via flight planned route, RENOB8B Departure, initial climb to 900 meters on QNH 1014, maintain 10100 meters on standard, squawk 3475.

P: Dongfang Delivery, say again all after QNH 1014.

C: VIR7937, maintain 10100 meters on standard, squawk 3475.

P: Cleared to Hong Kong via flight planned route, RENOB8B Departure, initial climb to 900 meters on QNH 1014, maintain 10100 meters on standard, squawk 3475, VIR7937.

C: VIR7937，可以经飞行计划航路飞往香港，RENOB8B 离场，起始爬升高度九百，修正海压幺洞幺四，在航路上保持标准气压幺洞幺巡航，应答机 3475。

P: 东方放行，请重复 QNH 幺洞幺四之后的信息。

C: VIR7937，在航路上保持标准气压幺洞幺巡航，应答机 3475。

P: 可以经飞行计划航路飞往香港，RENOB8B 离场，起始爬升高度九百，修正海压幺洞幺四，在航路上保持标准气压幺洞幺巡航，应答机 3475，VIR7937。

如果被呼叫单位不能确定谁呼叫自己，被呼叫单位可要求对方重复呼号直至建立联系。例如：

C: Station calling Beijing Ground, say again your call sign.

C: 哪个呼叫北京地面，请重复呼号。

如果管制员或驾驶员在发布指令或报告的过程中出现错误并立即更正，应使用"correction"（更正），重复更正后的正确部分。如果需要通过重复全部指令或报告才能更好地更正错误，可使用"correction I say again"（更正，我重复一遍）。例如：

C: BAW038, cleared to Beijing via flight planned route, NOMAD11D Departure, initial climb to 900 meters on QNH 1010, request level change en route, squawk 5310. Correction I say again, BAW038, cleared to Beijing via flight planned route, NOMAD11D Departure, initial climb to 900 meters on QNH 1014, request level change en route, squawk 5310.

P: Cleared to Beijing via flight planned route, NOMAD11D Departure, initial climb to 900 meters on QNH 1014, request level change en route, squawk 5310, BAW038.

C: BAW038, read-back correct.

C: BAW038，可以经飞行计划航路飞往北京，NOMAD11D 离场，起始爬升高度九百，修正海压幺洞幺洞，在航路上申请改变高度，应答机 5310。更正，我重复一遍，BAW038，可以经飞行计划航路飞往北京，NOMAD11D 离场，起始爬升高度九百，修正海压幺洞幺四，在航路上申请改变高度，应答机 5310。

P: 可以经飞行计划航路飞往北京，NOMAD11D 离场，起始爬升高度九百，修正海压幺洞幺四，在航路上申请改变高度，应答机 5310，BAW038。

C: BAW038，复诵正确。

当管制员或驾驶员认为对方接收可能有困难或有必要时，应重复通话中的重要内容，可使用"I say again"。例如：

P: Shanghai Approach, QFA301, 2700 meters maintaining, I say again, 2700 meters maintaining, engine losing power, engine losing power.

P: 上海进近，QFA301，两拐保持，我重复一遍，两拐保持，发动机失去推力，发动机失去推力。

练 习 题

说明：根据上下文完成通话。

P: Dongfang Delivery…1331.

C: _____, _____.

P: Dongfang Delivery, CCA1331.

C: _____, _____, _____.

P: Standing by, CCA1331.

　(A moment later)

C: CCA1331, Dongfang Delivery.

P: Dongfang Delivery, CCA1331, _____ Shanghai, Gate 5, request ATC clearance.

C: CCA1331, _____ you have got Information C.

P: _____, CCA1331.

C: CCA1331, _____ ATIS 128.2, call me when ready.

P: Monitoring 128.2, CCA1331.

　(A moment later)

P: Dongfang Delivery, CCA1331, _____ C, request ATC clearance.

C: CCA1331, Cleared to Shanghai via _____, ALPHA02 Departure, _____ to 900 meters on QNH 1011, maintain 9500 meters on standard, _____ 3475.

P: Cleared to Shanghai _____ flight planned route, ALPHA02 Departure, initial climb to 900 meters on QNH 1011, maintain 9500 meters on standard, squawk 3475, CCA1331.

C: CCA1331, _____ , _____ Dongfang Tower on 118.1.
P: 118.1, CCA1331.

1.6.2 许可的发布与复诵要求

驾驶员应向管制员复诵通过话音传送的 ATC 放行许可和指令中涉及安全的部分。具体来说，应复诵下列内容：

（1）空中交通管制航路放行许可；

（2）进跑道、起飞、着陆、滑行路线、穿越跑道、沿正在使用的跑道调头、跑道外等待的许可和指令。

（3）正在使用的跑道、高度表拨正值、二次监视雷达（SSR）编码、高度指令、航向与速度指令和空中交通管制员发布的或 ATIS 广播包含的过渡高度层。

对于其他许可和指令，包括附加条件许可，驾驶员复诵和认收的方式应能清楚表明自己已理解并将遵照执行这些许可和指令。

驾驶员应以呼号终止复诵。管制员必须监听驾驶员复诵。在肯定驾驶员复诵的内容正确时，可仅呼叫对方呼号。如果驾驶员复诵的指令或许可错误，管制员应明确发送"negative"（错误）后跟更正的内容。

如果对驾驶员能否遵照执行许可和指令有疑问，管制员在许可和指令后可加短语"if unable, advise"（如果不行，通知我），随后发布其他替换指令。任何时候，如果驾驶员认为不能遵照执行接收到的许可和指令，应使用短语"unable"（无法执行），并告知原因。

附加条件用语，如"在航空器着陆之后"或"在航空器起飞之后"，不应发布给对使用跑道有影响的活动，除非有关管制员或驾驶员能看见相关航空器或车辆。收到具有附加条件许可的航空器需要正确识别相关航空器或车辆。附加条件许可发布的格式如下：

（1）identification（识别标志）；

（2）the condition（条件）；

（3）the clearance（许可）；

（4）brief reiteration of the condition（条件的简要重复）。

例如：

C: CSS6878, behind the landing airbus on short final, line up behind.
P: Line up behind the landing airbus on short final, CSS6878.

C: CSS6878，在短五边上的空客落地后进跑道。
P: 在短五边上的空客落地后进跑道，CSS6878.

在发布应马上执行的指令，表明如果不执行指令将会造成严重的飞行冲突时，应使用"immediately"（立即）。在其他情况下，可使用"commence (action) now"（现在开始执行……）。

C: CES5186, go around immediately, aircraft on runway.

C: CES5186，立即复飞，跑道上有飞机。

练 习 题

说明：在空白处填上适当的内容，完成通话。

（1）**C:** DLH 723, _____.

　　　P: Turn left heading 130, DLH 723.

　　　C: DLH 723, negative, turn left heading 120.

　　　P: _____, DLH 723.

　　　C: DLH 723, read-back correct.

（2）**C:** HDA903, _____.

　　　P: Contact Dongfang Control 127.1.

　　　C: HDA903, negative, contact Dongfang Control on 127.0.

　　　P: _____.

　　　C: HDA903, read-back correct.

1.6.3　通信的移交

当航空器需要从一个无线电频率转换到另一个频率时，管制员应通知驾驶员转换频率。例如：

C: SIA5188, contact Dongfang Control on 127.5.
P: 127.5, SIA5188.

C: SIA5188，联系东方区域 127.5。
P: 127.5，SIA5188。

如果管制单位没有通知，驾驶员应在转换频率之前提醒管制员。例如：

P: Dongfang Control, YZR7968, request change to 127.5.
C: YZR7968, Dongfang Control, frequency change approved.

P: 东方区域，YZR7968，请求转换频率 127.5。
C: YZR7968，东方区域，同意转换频率。

当其他空中交通服务单位需要和航空器进一步通话时，可指示航空器 "stand by（frequency）"（在……频率上守听），此时管制单位应首先与航空器联系。例如：

C: KOR152, stand by for Dongfang Tower 118.1.
P: 118.1, KOR152.

C: KOR152，在东方塔台 118.1 上等待。
P: 118.1，KOR152。

指示航空器守听某广播频率时，应使用 "monitor（frequency）"。例如：

C: AIC348, Pudong Ground, monitor ATIS 127.85.
P: 127.85, AIC348.

C: AIC348，浦东地面，在 127.85 上守听通播。
P: 127.85，AIC348。

练　习　题

1. 听力练习

说明：听录音，填空。

（1）**C:** KLM570, contact ＿＿＿＿＿＿ on ＿＿＿. Correction ＿＿＿.
　　　P: ＿＿＿, KLM570.

（2）**C:** UAL781, contact ＿＿＿＿＿＿ on ＿＿＿. Correction ＿＿＿.
　　　P: ＿＿＿, UAL781.

2. 听力练习

说明：听录音，写出每句中的单位呼号及其联系频率，如下（1）所示。

（1）<u>Shanghai Control 135.7</u>＿＿＿.　　（2）＿＿＿＿＿＿＿＿＿＿＿.
（3）＿＿＿＿＿＿＿＿＿＿＿.　　　　（4）＿＿＿＿＿＿＿＿＿＿＿.
（5）＿＿＿＿＿＿＿＿＿＿＿.　　　　（6）＿＿＿＿＿＿＿＿＿＿＿.
（7）＿＿＿＿＿＿＿＿＿＿＿.　　　　（8）＿＿＿＿＿＿＿＿＿＿＿.
（9）＿＿＿＿＿＿＿＿＿＿＿.　　　　（10）＿＿＿＿＿＿＿＿＿＿＿.
（11）＿＿＿＿＿＿＿＿＿＿＿.

本章综合练习

一、听说练习

说明：请先听一遍下列数字组合的读法，听第二遍时请跟读。

5100 meters	280 knots
10100 meters	M0.85
11000 meters	121.45
FL 250	6565 kHz
QNH 1013	Runway 27
Heading 295	wind 350 degrees, 12 knots
Heading 360	2315

二、口语练习

说明：请按照发音规则读出下列数字和短语。

35007　　　　10100　　　　FL 330　　　　123.45　　　　QNH 1011

| 2500 | 324 feet | Runway 16 | 450 knots | RVR 600 |

三、口语练习

说明：角色扮演，两人一组。首先由驾驶员向管制员通报高度及到下一个报告点的时间，管制员要求航空器保持高度并在到达下一个报告点报告；然后交换角色再次练习。例如：

P: Beijing Control, CPA5852, maintaining 10100 meters, next report BHJ at 10.

C: Roger, CPA5852, maintain 10100 meters, report over BHJ.

P: Maintaining 10100 meters, wilco, CPA5852.

从表 1-31 中选取航空器呼号、高度、报告点和时间进行替换后练习。

表 1-31

航空器呼号	高　　度	报　告　点	时　　间
DKH1254	10400 meters	TIR	23
CSC8896	9500 meters	EGW	54
CHH7345	9200 meters	FXO	32
THY021	8900 meters	HAF	36
SWR197	7500 meters	SGK	07

四、听力练习

说明：听录音，把所听到的内容写出来。

（1） ＿＿＿＿＿＿＿＿＿＿＿＿＿＿＿.　　（2） ＿＿＿＿＿＿＿＿＿＿＿＿＿＿＿.

（3） ＿＿＿＿＿＿＿＿＿＿＿＿＿＿＿.　　（4） ＿＿＿＿＿＿＿＿＿＿＿＿＿＿＿.

（5） ＿＿＿＿＿＿＿＿＿＿＿＿＿＿＿.　　（6） ＿＿＿＿＿＿＿＿＿＿＿＿＿＿＿.

（7） ＿＿＿＿＿＿＿＿＿＿＿＿＿＿＿.　　（8） ＿＿＿＿＿＿＿＿＿＿＿＿＿＿＿.

（9） ＿＿＿＿＿＿＿＿＿＿＿＿＿＿＿.　　（10） ＿＿＿＿＿＿＿＿＿＿＿＿＿＿.

五、听力练习

说明：听录音，把所听到的句子写出来。

（1） ＿＿＿＿＿＿＿＿＿＿＿＿＿＿＿＿＿＿＿＿＿＿＿＿＿＿＿.

（2） ＿＿＿＿＿＿＿＿＿＿＿＿＿＿＿＿＿＿＿＿＿＿＿＿＿＿＿.

（3） ＿＿＿＿＿＿＿＿＿＿＿＿＿＿＿＿＿＿＿＿＿＿＿＿＿＿＿.

（4） ＿＿＿＿＿＿＿＿＿＿＿＿＿＿＿＿＿＿＿＿＿＿＿＿＿＿＿.

（5） ＿＿＿＿＿＿＿＿＿＿＿＿＿＿＿＿＿＿＿＿＿＿＿＿＿＿＿.

（6） ＿＿＿＿＿＿＿＿＿＿＿＿＿＿＿＿＿＿＿＿＿＿＿＿＿＿＿.

（7） ＿＿＿＿＿＿＿＿＿＿＿＿＿＿＿＿＿＿＿＿＿＿＿＿＿＿＿.

（8） ＿＿＿＿＿＿＿＿＿＿＿＿＿＿＿＿＿＿＿＿＿＿＿＿＿＿＿.

（9） ＿＿＿＿＿＿＿＿＿＿＿＿＿＿＿＿＿＿＿＿＿＿＿＿＿＿＿.

（10） ＿＿＿＿＿＿＿＿＿＿＿＿＿＿＿＿＿＿＿＿＿＿＿＿＿＿.

六、故事复述

说明：听录音，并复述故事内容。

七、小拓展：相似航班号

如果在同一波道有相似呼号，可能会出现混淆的情况。例如，管制员发给一架航空器的指令被另一架航空器接收并执行，这将造成冲突或事故。请思考：管制员如何从通话角度解决由于相似航班号引起驾驶员误听的情况？

_____

_____

_____

_____

_____

_____

第 2 章　机场管制——起飞前与起飞阶段

起飞前与起飞阶段主要包括航空器放行许可、推出、开车、滑行、进跑道和起飞等。此阶段所涉及的通话术语相对丰富、繁杂。特别是在停机位、机坪及滑行道上所使用的术语，往往要根据实际情景进行语言组织。此外，各类重要机场信息的通报也是此阶段术语中的重要内容。

通过本章的学习，应达到以下学习目标：

> ❖ 掌握无线电检查程序和用语；
> ❖ 掌握放行许可的格式和用语；
> ❖ 掌握推出、开车、滑行、进跑道等指令，以及起飞许可的格式和用语；
> ❖ 掌握重要信息通报的格式和内容；
> ❖ 理解离场条件的内容；
> ❖ 了解机场通播的各项内容及用语。

2.1　无线电检查

管制员或驾驶员认为必要时，可利用无线电检查程序，检查其无线电设备是否工作正常。

2.1.1　单词与词组

readability	清晰程度
transmission	发话
transmitter	发射机
box	无线电收发机
loud and clear	声音洪亮清楚
cut in and out	时断时续
loud background whistle	背景音刺耳
short count	短数
long call	长呼

2.1.2　听录音 🔊

2.1.3　典型格式与范例

> **1. 无线电检查程序应采用的形式**
>
> （1）对方电台呼号；
> （2）己方电台呼号；
> （3）无线电检查（radio check）；
> （4）使用的频率。
>
> **2. 无线电检查回答应按照的形式**
>
> （1）对方电台呼号；
> （2）己方电台呼号；
> （3）所发射信号的质量（readability）。

发射信号的质量如表 2-1 所示。

表 2-1[①]

信 号 质 量	英 文 描 述	中 文 描 述
unreadable（不清楚）	WUN	一个
readable now and then（可断续听到）	TOO	两个
readable but with difficulty（能听清但很困难）	TREE	三个
readable（清楚）	FOW-er	四个
perfectly readable（非常清晰）	FIFE	五个

信号检查质量的英文通话按照 1.3 节数字的标准发音读出；中文通话按照"信号一（两、三、四、五）个"读出。

范例：

（1）　**P:** Dongfang Ground, ABW422, **radio check on 121.5. How do you read me?**
　　C: ABW422, Dongfang Ground, **I read you 5.**

　　P: 东方地面，ABW422，无线电检查 121.5，你听我几个？
　　C: ABW422，东方地面，我听你五个。

（2）　**P:** Beijing Tower, KZR888, radio check on 118.55.
　　C: Station calling Beijing Tower, say again your call sign, you are unreadable.
　　（or）
　　C: KZR888, Beijing Tower, **reading you 3**, loud background whistle.

　　P: 北京塔台，KZR888，无线电检查 118.55。
　　C: 哪个呼叫北京塔台，重复一遍你的呼号，信号不清楚。
　　（或）
　　C: KZR888，北京塔台，听你三个，背景音刺耳。

---

① 此表参考民用航空行业标准《空中交通无线电通话用语》（MH/T 4014—2003）。

练 习 题

1. 听说练习

说明：听录音，把通话补充完整。听第二遍时请跟读。

（1）**P:** Dongfang Ground, CCA1356, radio check _____.

    C: CCA1356, Dongfang Ground, _____, your signal is _____, check your _____.

    (A moment later)

    P: Dongfang Ground, CCA1356,1, 2, 3, 4, 5,_____.

    C: CCA1356, _____.

（2）**P:** Dongfang Ground, SIA802, radio check _____. How do you read me?

    C: Station calling Ground, _____, check your transmitter and give me a short count.

（3）**P:** Dalian Ground, CQH631, radio check _____ on _____.

    C: CQH631,Dalian Ground, readability _____, _____.

2. 口语练习

说明：角色扮演，两人一组。在航空器离场前，驾驶员向管制员请求无线电检查，管制员给予回答。然后交换角色再次练习。例如：

> **P:** Dongfang Tower, <u>KAL332</u>, radio check <u>121.2</u>.
>
> **C:** <u>KAL332</u>, Dongfang Tower, readability 5.

从表 2-2 中选取航空器呼号和频率进行替换后练习。

表 2-2

航空器呼号	频　率
DLH721	118.5
FIN052	121.4
DAL185	123.5
ACA059	119.7
SIA802	118.9

3. 完成通话

说明：根据中文通话内容，写出对应的英文通话。

P: 东方地面，KLM889，无线电检查 121.6。

_____.

C: 哪个呼叫东方地面，重复一遍你的呼号。

_____.

P: 东方地面，KLM889，无线电检查 121.6，你听我几个？

C: KLM889，信号时断时续。

P: 东方地面，KLM889，5 分钟后再呼叫你。

（稍后）

P: 东方地面，KLM889，1，2，3，4，5，5，4，3，2，1，你现在听我几个？

C: KLM889，信号 5 个。

2.2 离场条件

2.2.1 离场信息

在航空器离场前，需要了解天气、使用跑道及高度表拨正值等信息，此类信息一般可由机场通播获取，如机场不提供机场通播，驾驶员可在请求开车之前向管制员请求离场条件。

1. 单词与词组

surface wind	地面风
gusting	阵风
wind calm	静风
wind variable	风向不定
dew point	露点
visibility	能见度
not available（not reported）	空缺（没有报告）
plus	正的
minus	负的
work in progress	正在施工
braking action	刹车效应

2. 听录音 🔊

3. 典型格式与范例

（1）航空器呼号（call sign）；
（2）使用跑道（runway-in-use）；
（3）风向、风速 (wind speed, wind direction)；
（4）高度表拨正值（QNH/QFE）；
（5）温度、露点（temperature, dew point）；
（6）能见度或跑道视程（visibility/RVR）。

范例：

P: Dongfang Ground, CQH345, IFR to Shanghai, request departure information.
C: CQH345, departure Runway 32, wind 290 degrees 4 knots, QNH 1022, temperature minus 2, dew point minus 3, RVR 550 meters, time 27.
P: Runway 32, QNH 1022, will call for start up, CQH345.

P: 东方地面，CQH345，仪表飞行规则，目的地上海，请求离场条件。
C: CQH345，离场跑道 32，地面风 290，4 节，修正海压 1 022，温度负 2，露点负 3，跑道视程 550 米，时间 27。
P: 跑道 32，修正海压 1 022，将请求开车，CQH345。

练 习 题

1. 听力练习

说明：读下列通话，先根据所学知识填空，再根据录音核对答案。

（1）P: Dongfang Tower, AFR496, _____ Paris, request _____ information.

　　C: AFR496, Dongfang Tower, _____, wind ____, ____ m/s, _____ 20 m/s, _____ more than 10 kilometers, temperature 25, dew point____, QNH_____.

　　P: Runway 18R, QNH_____, AFR496.

（2）P: Beijing Ground, SIA3310, _____.

　　C: SIA3310, Beijing Ground, _____, wind _____, _____ m/s, QNH 1011, temperature _____, dew point _____, RVR_____.

　　P: Runway 36L, QNH 1011, SIA3310.

（3）P: Hong Kong Ground, HDA101, request departure information.

　　C: HDA101, departure Runway_____, wind _____, _____ knots, _____ 25 knots, QNH _____, temperature _____, dew point 21, taxiway _____ closed due _____.

　　P: Roger, Runway 13, QNH 1023, HDA101.

2. 口语练习

说明：角色扮演，两人一组。在航空器离场前，驾驶员向管制员请求离场条件，管制员给予回答，驾驶员复诵。然后交换角色再次练习。例如：

P: Dongfang Tower, <u>BAW038</u>, request departure information.

C: <u>BAW038</u>, Dongfang Tower, Runway <u>36L</u>, wind 220 degrees, 12 m/s, QNH <u>1012</u>, temperature 23, dew point 20, RVR 1200 meters.

P: Runway <u>36L</u>, QNH <u>1012</u>, <u>BAW038</u>.

从表 2-3 中选取航空器呼号、跑道号和 QNH 值进行替换后练习。

表 2-3

航空器呼号	跑 道 号	QNH
CXA8460	36R	1010
OKA2876	18L	1011
CES2105	36L	1013
UAL889	18R	998
KLM898	36L	1014

2.2.2 机场通播

通播（ATIS）是在繁忙机场自动连续播放的情报服务，以减轻空中交通管制甚高频通信波道的通信负荷。ATIS 通常在一个单独的无线电频率上进行广播，一般每小时更新一次，天气变化迅速时也可随时更新，依次以字母代码 A，B，C，…，Z 表示，并按照 ICAO 标准读法读出。根据需要，通播可分为离场通播、进场通播和进离场通播。

1. 单词与词组

designator	代码
transition level	过渡高度层
transition altitude	过渡高度
ceiling	云底高
hectopascal	百帕
millibar	毫巴

2. 听录音

3. 典型格式与范例

（1）机场名称；
（2）通播代码；
（3）发布时间；

（4）使用跑道；

（5）重要的跑道道面情况[①]；

（6）地面风向风速；

（7）能见度、跑道视程；

（8）现行天气报告；

（9）大气温度、露点、高度表拨正值；

（10）趋势型着陆天气预报；

（11）其他必要的飞行情报及自动情报服务的信息。

范例：

Beijing capital international airport Information K. 0700UTC. Main landing Runway 36R, ILS approach, main departure Runway 36L. Wind 280 degrees, 6 m/s, gusting to 12 m/s. Visibility 4000 meters, intermittent light rain, overcast, ceiling 900 meters. Temperature 23, dew point 22. QNH 1006. Taxiway L closed. Advise on initial contact you have Information K.

北京首都国际机场情报通播 K，0700 协调世界时，主着陆使用跑道 36 右，ILS 进近，主起飞跑道 36 左。风向 280，6 米秒，阵风 12 米秒。能见度 4 000 米，间断小雨，阴天，云底高 900 米。温度 23，露点 22。修正海压 1006。滑行道 L 关闭。首次与管制员联系时报告您已收到通播 K。

练 习 题

1. 词汇练习

说明：将下列各组词按照程度从低到高排序，并指出该组词描述的是机场通播中的哪一项。

（1）moderate, light, heavy.

（2）wet, flooded, standing water, damp.

（3）good, poor, medium, medium to poor, good to medium, less than poor.

（4）scattered, few, overcast, sky clear, broken.

2. 听力练习

说明：听录音，填空。

This is Heathrow departure information _____. _____ hours, weather: _____, _____ knots, temperature _____, dew point _____. QNH _____, departure runway _____.

3. 听力练习

说明：听录音，找出下列信息。

ATIS_____, Time_____, Runway_____,

---

① 2021 年 11 月 4 日起，我国开始按照 ICAO 跑道表面状况全球通报格式（GRF）实施新的跑道表面状况评估和通报程序，具体通报要求可参见民航局相关文件。

SID_____, TL_____.

4. 听力练习

说明：听录音，圈出表 2-4 中正确的信息。

表 2-4

信 息 项	选 项			
ATIS designator	O	A	R	B
wind direction	310 degrees	360 degrees	160 degrees	60 degrees
wind speed	10～15 knots	15～35 knots	15～25 knots	25～35 knots
cloud amount	scattered	few	broken	overcast
cloud height	1000 feet	2000 feet	3000 feet	4000 feet
temperature	−1	−11	21	11
QNH	1004	1114	1104	1040

2.3 重要机场情报

　　重要机场情报是有关活动区及其相关设施的信息，这些信息对航空器的安全运行具有十分重要的意义。在航空器开车前或滑行前，以及开始最后进近前，除非已知航空器已从其他来源获取部分或全部的机场情报，否则，应尽可能向其提供重要机场情报。重要机场情报包括：活动区内及其邻近区域内进行的施工或维修工程；跑道上、滑行道上或停机坪上道面不平或破损情况（不管是否已做出标记）；跑道上、滑行道上或停机坪上有水、雪、雪浆、冰或霜的情况及其附近雪堆的情况；其他临时危害，包括地面上及空中的鸟类；机场灯光系统部分或全部不工作或故障情况；其他相关信息。

2.3.1　单词与词组

construction work	施工
work in progress	正在施工
slush	雪浆
snow bank	雪堆
unserviceable/inoperative/unavailable	不工作的、无用的
flock of birds	鸟群
water patches	块状积水
vehicle	车辆
wind shear	风切变

2.3.2　听录音🔊

2.3.3　范例

（1）　C: DKH1022, **caution construction work** adjacent to Gate 37.

C: DKH1022，注意 37 号桥附近有施工。

（2）　C: SBI874, **caution work in progress** ahead south side of Taxiway A.

C: SBI874，注意前方滑行道 A 南侧正在施工。

（3）　C: TBA9815, caution taxiway centreline **lighting unserviceable.**

C: TBA9815，注意滑行道中线灯不可用。

（4）　C: EPA6218, caution VASIS Runway 27 unserviceable.

C: EPA6218，注意 27 号跑道 VASIS 系统不可用。

（5）　C: CSC9988, caution large **flock of birds** north of Runway 27 near main taxiway.

C: CSC9988，注意 27 号跑道北侧主滑附近有大群鸟活动。

（6）　C: CHH7265, caution ILS Runway 09 unserviceable.

C: CHH7265，注意 09 号跑道 ILS 不可用。

（7）　C: **BAW038, threshold of runway 36 displaced 150 meters due broken surface.**

C: BAW038，由于道面破损，跑道 36 入口内移 150 米。

（8）　C: CAL3201, runway 36, surface condition code 2,5,5 issued at 1320. All parts slush, 5 millimeters, 100 per cent. ①

C: CAL3201，1320 发布的跑道 36 表面状况代码 2，5，5。所有部分雪浆，5 毫米，百分之百。

练 习 题

1. 词汇练习

说明：根据下列单词的发音，将其放入适当的位置。

available　　towing　　hangar　　failure　　incident　　information
intersection　reported　retracted　terminate　unserviceable

●●	●●●	●●●	●●●●	●●●●

2. 听力练习

说明：听录音，填空。

---

① 跑道表面状况具体包括跑道状况代码（RWYCC），以及跑道表面污染物的种类、深度和覆盖范围等。跑道状况代码，是一组从 6 到 0 的整数，可以直接表示跑道表面状况对飞机着陆和起飞滑跑性能的影响，有助于飞行机组对航空器的运行性能进行计算。

（1）**C:** AAL241, taxi with caution, _____ on Taxiway 2.

　　P: Roger, AAL241.

（2）**C:** MAS3351, taxi to holding point _____ . Be advised _____

partly covered by ice. Braking action_____ .

　　P: Roger. _____ , MAS3351.

（3）**C:** CHH801, taxi to holding point _____ . _____on taxiway

_____ .

　　P: Roger. _____ , CHH801.

3. 完成通话

说明：根据中文提示，写出相应的英文通话内容。

（1）**C:** CSN1233, _____ .（注意停机位 36

附近的施工。）

　　P: Roger, CSN1233.

（2）**C:** KLM898, _____ .（注意滑行道 A 的

北部正在施工，以红色旗子标出。）

　　P: Roger, KLM898.

（3）**C:** SAS995, _____ .（注意滑行道 A 旁

边的施工车辆。）

　　P: Roger, SAS995.

（4）**C:** CDG332, _____ .（注意滑行道 B 中

线灯光不能提供服务。）

　　P: Roger, CDG332.

（5）**C:** LKE278, wind calm, Runway 09, cleared for take-off. _____ .

（注意机场以东 1 千米有鸟群。）

　　P: Runway 09, cleared for take-off, LKE278.

2.4 放行许可

驾驶员在收到离场条件或离场 ATIS 之后，通常在开车前 5～10 分钟向地面或放行单位
请求放行许可。

当驾驶员需要记录，同时为避免无谓的重复，管制员应缓慢地、清楚地发布许可。放行
许可宜在开车前发布给驾驶员，不应在驾驶员对正跑道和实施起飞动作时发布放行许可。

放行许可抄收好后，驾驶员必须复诵许可内容，在被证实准确无误后请求推出开车。

2.4.1 单词与词组

flight planned route	计划航路
valid	有效
squawk	（调置）应答机
en route	航路上

2.4.2 听录音 🔊

2.4.3 典型格式与范例

（1）航空器呼号；
（2）许可界限（通常指目的地机场）；
（3）飞行的航路或航线；
（4）指定的标准仪表离场代号（没有代号的情况除外）；
（5）飞行高度；
（6）应答机编码；
（7）其他必要指令或信息，如转频的指令。

范例：

（1）**P:** Pudong Delivery, CCA102, destination Beijing with Information B, request ATC clearance.
C: CCA102, Information B is valid, **cleared to Beijing via flight planned route**, runway-in-use 17, PIKAS11D, **initial climb to 900 meters on QNH 1011, request level change for 9800 meters en route**, squawk 5310. Contact Approach on 123.8 when airborne.
P: Cleared to Beijing via flight planned route, runway-in-use 17, PIKAS11D, initial climb to 900 meters on QNH 1011, request level change for 9800 meters en route, squawk 5310, contact Approach on 123.8 when airborne, CCA102.
C: CCA102 read-back correct, contact Ground on 121.65, good day.

P: 浦东放行，CCA102，目的地北京，通播 B，请求放行许可。
C: CCA102，通播 B 有效，可以沿计划航路放行至北京，使用跑道 17，PIKAS11D 离场，起始爬升高度修正海压 900 米，修正海压 1 011，航路上申请巡航高度层 9 800 米，应答机 5310，离地后联系进近 123.8。
P: 可以沿计划航路放行至北京，使用跑道 17，PIKAS11D 离场，起始高度修正海压 900 米，修正海压 1 011，航路上申请巡航高度层 9 800 米，应答机 5310，离地后联系进近 123.8，CCA102。
C: CCA102，复诵正确。联系地面 121.65，再见。

（2）**P:** Dalian Ground, AAR3365, destination Seoul with Information G, request ATC clearance.
C: AAR3365, Information G is valid, cleared to Seoul via flight planned route, **initial altitude** 900 meters on QNH 998, ECH12D, **cruising level** 9800 meters, runway-in-use 28, squawk 6312. **After departure contact Approach on 123.3**.
P: Cleared to Seoul via flight planned route, initial altitude 900 meters on QNH 998, ECH12D, cruising level 9800 meters, runway-in-use 28, squawk 6312. After

P: 大连地面，AAR3365，目的地首尔，通播 G，请求放行许可。
C: AAR3365，通播 G 有效，可以沿计划航路放行至首尔，起始高度修正海压 900 米，修正海压 998，ECH12D 离场，巡航高度层 9 800 米，使用跑道 28，应答机 6312，离地后联系进近 123.3。
P: 可以沿计划航路放行至首尔，起始高度修正海压 900 米，修正海压 998，ECH12D 离场，巡航高度层 9 800 米，使用跑道 28，应答机 6312，离地后联系进近 123.3，

departure contact Approach on 123.3, AAR3365.

C: AAR3365, read-back correct.

AAR3365。

C: AAR3365，复诵正确。

练 习 题

1. 听力练习

说明：听录音，完成通话。

（1）**P:** Pudong Delivery, CCA1502, Gate 12 to Beijing, request ATC clearance.

　　C: CCA1502, Pudong Delivery, cleared to Beijing, PIKAS2D Departure,

　　_____, _____, _____.

（2）**P:** Pudong Delivery, CCA1506, Gate 12 to Beijing, request ATC clearance.

　　C: CCA1506, Pudong Delivery, cleared to Beijing, _____, PIKAS2X Departure,

　　_____, _____, _____.

2. 口语练习

说明：根据上下文填空。

P: Dongfang _____, D-YGTV.

C: _____, Dongfang Delivery.

P: Dongfang Delivery, D-YGTV, _____ 17, _____ Guangzhou, _____ C, request _____ clearance.

C: _____, _____ to Guangzhou _____, ALF11D Departure, initial climb to 600 meters on QNH 1011, maintain 9800 meters on standard, _____ 3475, after departure,_____Approach 128.1.

P: _____

_____, D-YGTV.

C: _____.

3. 口语练习

说明：根据图 2-1 和图 2-2 中的进程单及相关信息，编写管制员与驾驶员之间关于放行许可的对话。

（1）Information C, QNH 1010, Approach 123.8.

CSN3599		P/B		09/28　　1140	M
B737　M　　A 0514	A120	S/T	XJT-12D	8400	E
ZSQD 1200/1500 ZGGG		TAX		P12	A
		R/W　17		W/Z	I

图 2-1

（2）Information X, QNH 1009, Approach 123.3.

CCA1383		A090	P/B		09/28　1140		M
B737　M　A 2230			S/T　　　BRAVO01D			8400	E
ZTTT 1200/1330 ZSSS			TAX		P12		A
			R/W 11		W/Z		I

图 2-2

2.5　推出开车

通常情况下，航空器在停放时机头朝向候机楼，在离场时需利用推车将航空器推出停机位，这称为 push-back。少数航空器能利用自身动力推出，这称为 power-back。根据机场程序，驾驶员可向空管部门或机坪管制部门申请推出。

航空器开车前需要向管制员提出申请，有助于管制员做计划，避免航空器在地面等待时间过长，耗费过多燃油。航空器在提出开车申请时，应表明航空器位置和已接收的通播代号。

2.5.1　单词与词组

push-back	推出
at own discretion	自己决定
slot time	时隙（航空器预计离场时间段）
stand	停机位
gate	登机门，停机位
stand by	稍等
start up	开车（启动发动机）
breakdown	故障

2.5.2　听录音 🔊

2.5.3　范例

（1）　P: Dongfang Tower, GIA893, Stand 27, request push-back.
C: GIA893, **push-back approved, facing east**.

P: 东方塔台，GIA893，停机位 27，请求推出。
C: GIA893，同意推出，机头朝东。

（2）　P: Dongfang Tower, KLM4302, Stand 27, request push-back.
C: KLM4302, stand by. **Expect one minute delay due** B747 taxiing behind.

P: 东方塔台，KLM4302，停机位 27，请求推出。
C: KLM4302，稍等。由于 B747 在后面滑行，预计延误一分钟。

（3）　P: Dongfang Tower, QFA129, Gate 24, request start up, Information B.
C: QFA129, **start up approved**.

P: 东方塔台，QFA129，停机位 24，请求开车，通播 B。
C: QFA129，同意开车。

（4）　P: Dongfang Tower, ANZ080, Stand 24, request start up, Information B.
C: ANZ080, **start up at 35**.

P: 东方塔台，ANZ080，停机位 24，请求开车，通播 B。
C: ANZ080，35 分开车。

（5）　P: Dongfang Tower, AAR3365, Stand 24, request start up, Information B.
C: AAR3365, expect start up time at 35.

P: 东方塔台，AAR3365，停机位 24，请求开车，通播 B。
C: AAR3365，预计开车 35 分。

（6）　P: Dongfang Tower, AFR381, Gate 24, request start up, Information B.
C: AFR381, expect departure 49, **start up at own discretion**.

P: 东方塔台，AFR381，停机位 24，请求开车，通播 B。
C: AFR381，预计离场 49 分，开车时间自己掌握。

（7）　C: CKK206, can you taxi in under your own power or do you want to be towed?

C: CKK206，能靠自身动力滑回来吗？还是需要拖车拖？

练　习　题

1. 完成通话

说明：假定你是管制员，根据英文提示完成通话。

（1）P: Dongfang Tower, CES5357, on apron, request push-back, Information B.
（Tell CES5357 there is a 5-minute delay because a B747 is taxiing behind CES5357.）
C: _____.
P: Roger, CES5357.

（2）P: Dongfang Tower, PIA852, Gate H, request start up, Information B.
（Tell PIA852 that the departure time is expected to be at 23 and it can start up at own discretion.）
C: _____.
P: Roger, PIA852.

（3）P: Dongfang Tower, SIA802, Stand 26, request push-back and start up, Information B.
（Permit the request.）
C: _____.
P: Roger, SIA802.

2. 口语练习

（1）说明：角色扮演，两人一组。首先由驾驶员向管制员申请开车，管制员给予回答。然后交换角色再次练习。例如：

P: Dongfang Tower, KLM898, Stand 27, request start up.

C: KLM898, start up approved.

从表 2-5 中选取航空器呼号、停机位号及管制员回答进行替换后练习。

表 2-5

航空器呼号	停机位号	管制员回答
P-TGUC	Stand 12	start up approved
JA-FRB	Gate 21	start up at own discretion
CSH9070	Gate 11	expect departure 13, start up at own discretion
G-NWTY	Stand 14	start up at 24
SIA802	Stand 23	expect start up at 45

（2）说明：角色扮演，两人一组。首先由驾驶员向管制员申请推出，由于某些原因，管制员需要发布延迟推出的指令。然后交换角色再次练习。例如：

> **P:** Dongfang Ground, CCA191, ready for push-back.
>
> **C:** CCA191, expect <u>one minute delay</u> due <u>Airbus taxiing behind</u>.

从表 2-6 中选取预计延误时间和原因进行替换后练习。

表 2-6

预计延误时间	原　　因
5 minutes delay	traffic
several minutes delay	vehicle breakdown behind you
push-back in about an hour	flow control

3. 听力练习

说明：听录音，填空。

（1）**P:** Beijing Ground, CXA505, Stand _____, request start up.

　　C: CXA505, _____.

（2）**P:** Beijing Ground, CSH100, Stand_____, request start up and pushback. Information _____. Our slot-time is _____ plus 6 minutes.

　　C: CSH100, start up and push at _____.

（3）**P:** Beijing Ground, AFR710 _____, request start up and pushback.

　　C: AFR710, expect departure at _____. Start up _____ and pushback _____.

（4）**P:** Beijing Ground, JAL410, _____, request pushback.

　　C: JAL410, _____minutes delay due _____. _____.

4. 思考题

说明：管制员给航空器下达推出指令后，发现航空器后方有其他航空器滑过，应如何发指令以阻止冲突发生？

2.6 滑出

离场航空器准备就绪，在得到管制员的滑行指令后离开停机位，滑行到起飞跑道外等待点。在一些大型机场，由于拥有多条跑道，以及跑道和滑行道的设计问题，导致航空器在滑行前往跑道的过程中，可能需要穿越另一条跑道。

2.6.1 单词与词组

taxi	滑行
taxi with caution	滑行时注意
taxi straight ahead	一直往前滑
expedite taxi	加速滑行
taxi slower	减速滑行
continue taxi	继续滑行
hold	等待
holding point	等待点
backtrack	（跑道上）调头
hold short of	在……外等待
hold position	原地等待
vacate	脱离
cross	穿越
give way to	给……让路
follow	跟在……后面
overtake	超越
traffic in sight	看到活动
hangar	机库
trench	壕沟

2.6.2 听录音 🔊

2.6.3 范例

（1） **P:** Dongfang Tower, LKE1912, Stand 27, request taxi.

P: 东方塔台，LKE1912，停机位 27，请求滑行。

C: LKE1912, Dongfang Tower, **taxi to holding point Runway** 18L **via Taxiway** H3, A6 and B, **hold short of** Runway 18L.

C: LKE1912，东方塔台，沿滑行道 H3、A6、B 滑到 18 左跑道等待点，在 18 左跑道外等待。

（2） **P:** Dongfang Tower, GDC7128, heavy, Stand 27, request taxi, Information C.

P: 东方塔台，GDC7128，重型，停机位 27，请求滑行，通播 C。

C: GDC7128, Dongfang Tower, taxi to holding point Runway 27, **give way to** B747 passing left to right.

C: GDC7128，东方塔台，滑到 27 号跑道等待点，给从左向右滑行的 B747 让路。

（3）P: Dongfang Tower, DAL185, Stand 36, **request backtrack** at present position.

P: 东方塔台，DAL185，停机位 36，请求原地调头。

C: DAL185, Dongfang Tower, **backtrack approved**.

C: DAL185，东方塔台，同意原地调头。

（4）P: Dongfang Tower, CES5301, approaching holding point, **request crossing Runway 22**.

P: 东方塔台，CES5301，接近等待点，请求穿越 22 号跑道。

C: CES5301, **cross Runway 22**, **report vacated**.

C: CES5301，穿越 22 号跑道，脱离报告。

P: CES5301, crossing.

P: CES5301，正在穿跑道。

（A moment later）

（稍后）

P: CES5301, **runway vacated**.

P: CES5301，已脱离跑道。

练 习 题

1. 听说练习

说明：听录音，填空。听第二遍时请跟读。

（1）P: Pudong Ground, CCA981, request taxi.

C: CCA981, _____ via Taxiway _____ and _____.

P: _____ via Taxiway _____ and _____, CCA981.

（2）P: Ground, ANA211, Stand ___, request taxi to holding point Runway ____.

C: ANA211, taxi to holding point ____ Runway ____.

（3）P: _____, ACA898, Stand D2, request detailed taxi instructions for Runway ____.

C: ACA898, taxi via Taxiway ____ and _____. Taxi _____, marked _____ taxiway at holding point. Give way to Boeing 737 entering _____ from taxiway.

P: Traffic _____, ACA898.

（4）C: UAL451, hold short of taxiway ____.

P: Holding short, UAL451.

2. 听说练习

说明：听录音，写句子。听第二遍时请跟读。

（1）_____.

（2）_____.

（3）_____.

（4）_____.

（5）_____.

（6）_____.

（7）_____.

（8）_____.

3. 口语练习

说明：根据图 2-3 中所示的各种情况，向航空器发出合适的指令。例如：

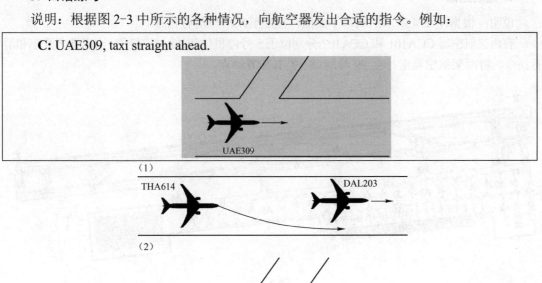

C: UAE309, taxi straight ahead.

（1）

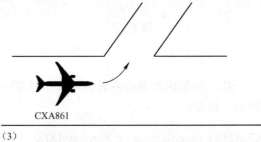

（2）

（3）

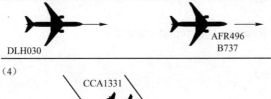

（4）

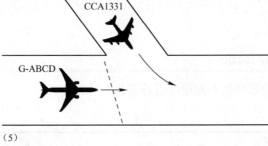

（5）

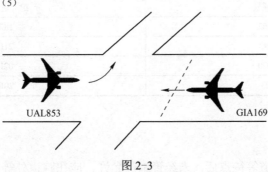

图 2-3

4. 完成通话

说明：根据图 2-4 所示的机场平面图，按照要求给出滑行指令。

有两架航空器 CCA101 和 CCA102 分别位于 5 号停机位和 40 号停机位，给出推出开车和滑行指令，将两架航空器滑行至 29 号跑道 I 类 ILS 等待点。

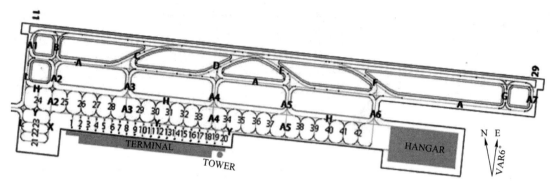

图 2-4

5. 口语练习

说明：角色扮演，两人一组。首先由航空器驾驶员向管制员申请穿越跑道，管制员给予回答。然后交换角色再次练习。例如：

P: Dongfang Tower, CES2105, request crossing Runway 18L .

C: CES2105, cross Runway 18L, report vacated.

P: CES2105, crossing.
（A moment later）

P: CES2105, runway vacated.

从表 2-7 中选取航空器呼号和跑道号进行替换后练习。

表 2-7

航空器呼号	跑道号
CSN3102	18R
HDA890	21L
CSH9070	02L
CPA903	20R

2.7　起飞

当航空器滑行至跑道等待点后，未经管制员允许，应在跑道外等待。管制员下达进入跑道指令后，航空器方可进入跑道。

在得到塔台管制员的起飞许可后，航空器开始起飞。起飞指航空器从起飞线开始滑行到离开地面，爬升到安全高度的加速运动过程。

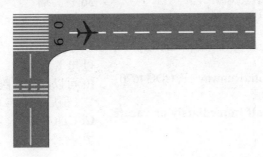

2.7.1 单词与词组

line up	进跑道
airborne	离地，升空
take-off	起飞
departure	离场
climb straight ahead	直线上升

2.7.2 听录音 🔊

2.7.3 范例

（1） **C:** DAL185, report ready for departure.
P: Wilco, DAL185.
（A moment later）
P: DAL185, ready.
C: DAL185, **Runway 09, line up and wait**.
P: Lining up, Runway 09, DAL185.
C: DAL185, **wind 360 degrees, 4 m/s, Runway 09, cleared for take-off**.
P: Runway 09, cleared for take-off, DAL185.

C: DAL185，准备好报。
P: 照办，DAL185。
（稍后）
P: DAL185，准备好了。
C: DAL185，进跑道 09 等待。
P: 进跑道，跑道 09，DAL185。
C: DAL185，地面风 360，4 米秒，跑道 09，可以起飞。
P: 跑道 09，可以起飞，DAL185。

（2） **C:** AAL186, report the Airbus on final in sight.
P: Airbus in sight, AAL186.
C: AAL186, **behind the landing traffic, line up and wait behind**.

P: After the Airbus line up and wait behind, AAL186.

C: AAL186，看到五边上的空客报告。
P: 看到空客了，AAL186。
C: AAL186，跟在落地飞机后面进跑道等待。
P: 跟在落地飞机后面进跑道等待，AAL186。

（3） **C:** CHH7378, **line up, be ready for immediate departure**.

C: CHH7378，进跑道，做好立即离场的准备。

P: Lining up, CHH7378.　　　　　　　　　P: 进跑道，CHH7378。

（4）　C: CXA4241, **advise able to depart from Runway** 36R **intersection** N.
P: Affirm, CXA4241.

C: CXA4241，能否接受道口 N 进跑道 36 右离场？
P: 是的，CXA4241。

（5）　C: CDG4670, Dongfang Tower, Runway 18, cleared for take-off.
P: Cleared for take-off, Runway 18, CDG4670.
（A moment later）
C: CDG4670, **take off immediately or vacate runway**①.

C: CDG4670，东方塔台，跑道 18，可以起飞。
P: 可以起飞，跑道 18，CDG4670。
（稍后）
C: CDG4670，立即起飞，否则脱离跑道。

（6）　C: CSN8875, after departure, **track extended centreline** 1500 meters before turning left.
P: Track extended centreline to 1500 meters before turning left, CSN8875.

C: CSN8875，起飞后，沿跑道中心延长线飞行到 1 500 米，然后左转。
P: 沿跑道中心延长线飞行到 1500，然后左转，CSN8875。

（7）　C: CPA347, **take off immediately or hold short of runway**.
P: Holding short, CPA347.

C: CPA347，立即起飞，否则跑道外等待。
P: 跑道外等待，CPA347。

（8）　C: EVA712, Dongfang Tower, Runway 18, cleared for take-off.
P: Cleared for take-off, Runway 18, EVA712.
（A moment later）
C: EVA712, **hold position, cancel take-off, I say again, cancel take-off**②.
P: Holding, EVA712.

C: EVA712，东方塔台，跑道 18，可以起飞。
P: 可以起飞，跑道 18，EVA712。
（稍后）
C: EVA712，原地等待，取消起飞，我重复一遍，取消起飞。
P: 原地等待，EVA712。

（9）　C: UAE309, Dongfang Tower, Runway 18, cleared for take-off.
P: Cleared for take-off, Runway 18, UAE309.
（UAE309 starts take-off roll.）
C: UAE309, **stop immediately**, UAE309, **stop immediately**.
P: Stopping, UAE309.

C: UAE309，东方塔台，跑道 18，可以起飞。
P: 可以起飞，跑道 18，UAE309。
（UAE309 开始起飞滑跑。）
C: UAE309，立即中断起飞，UAE309，立即中断起飞。
P: 中断起飞，UAE309。

练　习　题

1. 听力练习

说明：根据录音，完成管制员通话。

---

① 当起飞许可未被执行时。
② 航空器处于静止状态时取消起飞许可。

（1）**C:** CSN113, _____?

　　P: Ready, CSN113.

　　C: CSN113, _____.

　　P: Lining up, Runway 09, CSN113.

（2）**C:** JAL808, _____.

　　P: Ready for departure, JAL808.

　　C: JAL808, _____.

　　P: Behind Boeing 737 on final, lining up, JAL808.

　　C: JAL808, correct.

（3）**C:** SIA151, _____.

　　P: Cleared for take-off, Runway 23, wilco, SIA151.

（4）**C:** CES2014, _____.

　　P: Cleared for take-off, Runway 18, CES2014.

　　C: CES2014, _____.

　　P: Holding, CES2014.

（5）**C:** CES5130, _____.

　　P: Taking off, Runway 19, CES5130.

　　C: CES5130, _____.

　　P: Stopping, CES5130.

2．口语练习

（1）说明：角色扮演，两人一组。首先由驾驶员汇报航空器位置，管制员向其发布条件性进跑道许可。然后交换角色再次练习。例如：

> **P:** Dongfang Tower, <u>CCA101</u>, holding at holding point Runway <u>36L</u>.
>
> **C:** <u>CCA101</u>, hold short of runway, report <u>B747</u> on final in sight.
>
> **P:** <u>B747</u> in sight, <u>CCA101</u>.
>
> **C:** <u>CCA101</u> behind the landing aircraft, line up Runway <u>36L</u> behind.
>
> **P:** Behind the landing aircraft, line up behind, <u>CCA101</u>.

从表 2-8 中选取航空器呼号、跑道号及着陆航空器机型进行替换后练习。

表 2-8

航空器呼号	跑 道 号	着陆航空器机型
DLH620	29	DC-9
HDA270	13	A320
JAL782	03	B777
KAL930	21	B747

（2）说明：角色扮演，两人一组。首先由驾驶员向管制员报告准备好起飞，管制员给出起飞许可。然后交换角色再次练习。例如：

> **P:** Baiyun Tower, <u>CSN302</u>, ready for departure.
>
> **C:** <u>CSN302</u>, <u>wind 310 degrees, 15 knots</u>, Runway 23, cleared for take-off.
>
> **P:** Cleared for take-off, Runway 23, <u>CSN302</u>.
>   （A moment later）
>
> **C:** <u>CSN302</u>, take off immediately or vacate runway.
>
> **P:** Taking off, <u>CSN302</u>.

从表 2-9 中选取航空器呼号及风向风速进行替换后练习。

表 2-9

航空器呼号	风向、风速
F-ARFK	310 degrees, 15 m/s
D-RHLM	270 degrees, 8 m/s
G-PQNH	160 degrees, 12 m/s
N-SERD	190 degrees, 6 knots

3. 口语练习

说明：根据图 2-5 中所示的各种情景，向在跑道外等待点的航空器 CCA101 发出管制指令。

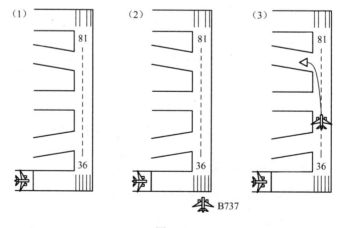

图 2-5

4. 思考题

（1）阅读下列两段对话，指出两段对话中管制员指令含义的区别。

① **C:** CCA101, wind calm, Runway 28, cleared for take-off.

C: CCA101, hold position, cancel take-off, I say again, cancel take-off due runway incursion.

　　P: Holding, CCA101.

② **C:** CCA101, wind calm, Runway 28, cleared for take-off.

　　C: CCA101, stop immediately, CCA101, stop immediately. Acknowledge.

　　P: Stopping, CCA101.

（2）指出上面的对话分别对应图 2-6 中的哪种情景。

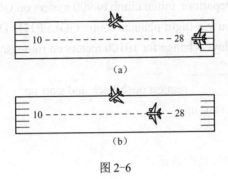

图 2-6

本章综合练习

一、听力练习

说明：听录音，填空。

P: Dongfang Tower…

C: _____, _____.

P: Dongfang Tower, G-JMLF, _____ 118.1.

C: G-JMLF, Dongfang Tower, _____ 2, _____ and _____ a _____.

P: Roger, G-JMLF.

　　（A moment later）

P: Dongfang Tower, G-JMLF, 1, 2, 3, 4, 5, how do you read me?

C: _____, _____ 5.

P: Thank you, G-LF.

P: Dongfang Tower, G-LF, destination Guangzhou, Stand 26, request ATC clearance.

C: G-LF, _____ you have got Information C.

P: Negative, G-LF.

C: G-LF, _____ ATIS 128.2, call me when ready.

P: Monitoring 128.2, G-LF.

　　（A moment later）

P: Dongfang Tower, G-LF, Information C, ready to start.

C: G-LF, Information C is valid, _____ to Guangzhou via flight planned route, GOLFF11D Departure, initial climb to 900 meters on QNH 997, request level change for 10100 meters en route, squawk 5246. _____, _____, G-LF, cleared to Guangzhou via flight planned route, GOLFF11D Departure, initial climb to 900 meters on QNH 998, request level change for 10100 meters en route, squawk 5246.

P: Dongfang Tower, G-LF, _____ request level change for 10,100 meters en route.

C: G-LF, GOLFF11D Departure, initial climb to 900 meters on QNH 998.

P: Cleared to Guangzhou via flight planned route, GOLFF11D Departure, initial climb to 900 meters on QNH 998, request level change for 10100 meters en route, squawk 5246, G-LF.

C: _____, _____.

P: _____, _____, request push-back and start up.

C: G-LF, push-back and start up _____.

二、听力练习

说明：听录音，完成管制员的通话。

C1: _____.

P: Taxi via H, Z, N to holding point, Runway 35R, CCA981.

（A moment later）

C1: _____.

P: Tower 118.1, CCA981.

P: Pudong Tower, CCA981, approaching holding point Runway 35R.

C2: _____.

P: Holding short, CCA981.

（A moment later）

P: CCA981, B737 in sight.

C2: _____.

P: Behind B737 on final, line up behind, CCA981.

C2: _____.

P: CCA981, lining up.

C2: _____.

P: CCA981, ready.

C2: _____.

P: Cleared for take-off, Runway 35R, will report airborne CCA981.

（A moment later）

P: CCA981, airborne 05.

C2: _____.

P: 118.8, good day CCA981.

三、口语练习

说明：根据英文提示，写出陆空通话指令。

P: Wuhan Ground, CES521, Destination Chengdu, with Information G, request push-back and start up.

（You notice that a B737 is taxiing behind CES521. Tell CES521 that the request is not permitted.）

C:_____.

P: Roger, CES521.

（Permit CES521 to push-back and start up, nose to the north.）

C: _____.

　（Read-back omitted）

P: Wuhan Ground, CES521, ready for taxi.

（Give CES521 taxi instructions. Taxiways are A6 and A, and the clearance limit is the holding point of Runway 36R.）

C: _____.

　（Read-back omitted）

（You notice that there is snow and ice on some part of Taxiway C. Alert CES521 to taxi with care.）

C: _____.

四、听力练习

说明：图 2-7 中所示的滑行道上有八架飞机，分别用字母 A 至 H 标出。听录音，判断听到的指令针对哪一架飞机。

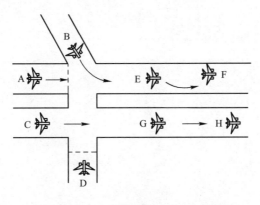

图 2-7

（1）_____；　（2）_____；

（3）_____；　（4）_____；

（5）_____.

五、故事复述

说明：听录音，并复述故事内容。

六、小拓展：监听复诵

航空器驾驶员复诵不正确怎么办？你将听到一段有关地面滑行的实况录音。听录音并回答以下问题。

（1）为什么监听航空器驾驶员的复诵很重要？

_____

_____

_____ .

（2）管制员第一次给出的滑行指令是：

_____

_____

_____ .

　　航空器驾驶员的复诵是：

_____

_____

_____ .

（3）当发现航空器驾驶员复诵错误时，管制员采取了哪些措施？

_____

_____

_____ .

第 3 章　进近管制——离场阶段

航空器起飞后，塔台管制员将其移交给进近管制员指挥。在非常繁忙的终端管制区，进近管制可划分为不同的管制席位，各自承担不同的管制职责，如离场雷达管制、进场雷达管制和五边雷达管制等。在提供空中交通监视（雷达）服务时，管制员应使用一般雷达管制用语、二次监视雷达用语和进近雷达管制用语等雷达管制标准术语。

对于 IFR 离场的航空器而言，管制员的指令主要涉及高度的改变和路径的选择两方面。高度的改变以上升指令为主，高度保持指令为辅；而路径的选择则以指定离场程序（标准仪表离场程序或 RNAV 离场程序）或具体航向、航迹的方式体现。除了常规指令以外，当航空器之间存在飞行冲突或其他必要情况时，通常要进行活动通报。

通过本章的学习，应达到以下学习目标：

❖ 掌握雷达管制基本用语；
❖ 掌握 IFR 航空器离场的用语；
❖ 掌握飞行活动通报的相关用语；
❖ 理解飞行活动冲突通报的时机及内容。

3.1　雷达管制用语

雷达管制用语主要包括一般雷达管制用语、进近雷达管制用语和二次监视雷达用语。一般雷达管制用语既可用于一次监视雷达设备，也可用于二次监视雷达设备提供的管制服务，而二次监视雷达用语只适用于航空器配备机载应答机的二次监视雷达管制服务。进近雷达管制用语是进近管制员指挥航空器进场和进近时的用语。本节主要介绍一般雷达管制用语和二次监视雷达用语。进近雷达管制用语将在第 5 章进行介绍。

3.1.1　单词与词组

identification	识别
radar contact	雷达看到
identified	已识别
resume own navigation	恢复自主领航
direct	直飞
magnetic track	磁航迹
remain this frequency	保持长守
terminate	终止
orbit	盘旋
transponder	应答机
confirm transponder operating	证实应答机在工作

radio contact lost	失去无线电联络
observed	看到了
if you read	如果你能听到
low altitude warning	低高度告警
minimum flight altitude	最低飞行高度
terrain alert	近地告警

3.1.2　听录音 🔊

3.1.3　范例

（1）　C: THY021, **report heading and level**.
P: Heading 140, maintaining 3000 meters, THY021.
C: THY021, **for identification, turn left heading 110**.
P: Left heading 110, THY021.
（A moment later）
P: Wuhan Approach, THY021, on heading 110.
C: THY021, **identified, position 20 kilometers north of CON, continue present heading**.
P: Continue present heading, THY021.

　C: THY021，报告航向、高度。
P: 航向 140，高度 3 000 米保持，THY021。
C: THY021，为了识别，左转航向 110。

P: 左转航向 110，THY021。
（稍后）
P: 武汉进近，THY021，航向 110。

C: THY021，雷达已识别，位置从化以北 20 千米，保持当前航向。
P: 保持现在航向，THY021。

（2）　C: DLH723, **not identified**. Not yet within radar coverage.

　C: DLH723，没有识别。还未到雷达覆盖范围内。

（3）　C: CSZ9783, **identification lost due radar failure**. Contact Dongfang Control on 126.25.
P: 126.25, CSZ9783.

　C: CSZ9783，由于雷达失效，识别丢失，联系东方区域 126.25。
P: 126.25，CSZ9783。

（4）　C: CCA3653, **will shortly lose identification due blind area**. **Remain this frequency**.
P: Wilco, CCA3653.

　C: CCA3653，由于雷达盲区，将短时失去雷达识别，保持长守。
P: 照办，CCA3653。

（5）　C: HBH8380, **turn right heading 350 for spacing**.
P: Right heading 350, HBH8380.
C: HBH8380, **stop turn, heading 290**.
P: Heading 290, HBH8380.

　C: HBH8380，由于间隔，右转航向 350。

P: 右转航向 350，HBH8380。
C: HBH8380，停止转弯，航向 290。
P: 航向 290，HBH8380。

（6）　C: CSC8861, **after HO, fly heading 010**.
P: After HO, fly heading 010, CSC8861.

　C: CSC8861，过长武后，航向飞 010。
P: 过长武后，航向飞 010，CSC8861。

（7） C: CSN9586, report heading.   C: CSN9586，报告航向。

P: Heading 240, CSN9586.   P: 航向 240，CSN9586。

C: CSN9586, **continue heading 240**.   C: CSN9586，保持航向 240。

P: Continue heading 240, CSN9586.   P: 保持航向 240，CSN9586。

（8） C: CHH345, **make a three sixty turn left for sequencing**.   C: CHH345，由于排序，左转一圈。

P: Three sixty turn left, CHH345.   P: 左转一圈，CHH345。

（9） C: CXA8470, **orbit left for delay**.   C: CXA8470，由于延迟，左转盘旋。

P: Orbit left, CXA8470   P: 左转盘旋，CXA8470。

（10） C: ACA050, **resume own navigation to DAL**[①], position 15 kilometers north of DAL.   C: ACA050，恢复自主领航，飞往大理，位置大理以北 15 千米。

P: Direct DAL, resume own navigation, ACA050.   P: 飞往大理，恢复自主领航，ACA050。

（11） C: GCR7896, **position 12 kilometers from touchdown, radar service terminated**, contact Tower 118.1.   C: GCR7896，位置距接地点 12 千米，雷达服务终止，联系塔台 118.1。

P: 118.1, GCR7896.   P: 118.1，GCR7896。

（12） C: UAL326, **advise transponder capability**.   C: UAL326，报告应答机能力。

P: **Transponder Charlie**, UAL326.   P: 模式 C 应答机，UAL326。

C: UAL326, **squawk** 6411.   C: UAL326，应答机 6411。

P: 6411, UAL326.   P: 6411，UAL326。

C: UAL326, **confirm squawk**.   C: UAL326，证实应答机。

P: Squawking 6411, UAL326.   P: 应答机 6411，UAL326。

C: UAL326, **reset** squawk 6411.   C: UAL326，重置应答机 6411。

P: Resetting 6411, UAL326.   P: 重置应答机 6411，UAL326。

C: UAL326, radar contact. Climb to 4500 meters.   C: UAL326，雷达看到，上升到 4 500 米。

P: Climbing to 4500 meters, UAL326.   P: 上升到 4 500 米，UAL326。

（13） C: JAL345, **check altimeter setting and confirm level**.   C: JAL345，检查高度表设定值，证实高度。

P: **Altimeter** 1003, 3600 meters, JAL345.   P: 修正海压 1003，高度 3 600 米，JAL345。

C: JAL345, **stop squawk Charlie, wrong indication**.   C: JAL345，关闭应答机 C 模式，指示错误。

（14） C: B3475, **radio contact lost. If you read, squawk IDENT, I say again, squawk IDENT**.   C: B3475，失去无线电联络，如果你能听到，应答机识别，我重复一遍，应答机识别。

（A moment later）   （稍后）

---

① 此处来自 ICAO 2007 版 9432 号文件 *Manual of Radiotelephony*。

C: B3475, **squawk observed, will continue radar control, squawk IDENT to acknowledge**.

C: B3475，应答机识别看到了，将继续雷达管制，应答机识别进行认收。

（15）**C:** CHB6237, **reply not received**. If you read, turn left heading 150, I say again, turn left heading 150.

（A moment later）

C: CHB6237, **turn observed**, position 15 kilometers south of VYK VOR, will continue radar control.

C: CHB6237，没有收到回答，如果你能听到，左转航向 150，我重复一遍，左转航向 150。

（稍后）

C: CHB6237，转弯看到了，位置大王庄以南 15 千米，将继续雷达管制。

（16）**C:** CUA5605, **radar service terminated due radar failure**, **converting to procedural control** (or **non-radar separation**).

P: Roger, CUA5605.

C: CUA5605，雷达服务终止，由于雷达失效，转换到程序管制（非雷达间隔）。

P: 收到，CUA5605。

（17）**C:** TBA9907, **low altitude warning**, check your altitude immediately, QNH 1006, minimum flight altitude 650 meters.

P: Roger, TBA9907.

C: TBA9907，低高度告警，立即检查高度，修正海压 1 006，最低飞行高度 650 米。

P: 收到，TBA9907。

（18）**C:** CSN6580, **terrain alert**, climb to 900 meters, QNH 1006.

P: Climbing to 900 meters, QNH 1006, CSN6580.

C: CSN6580，近地告警，上升到修正海压 900 米，修正海压 1 006。

P: 上升到修正海压 900 米，修正海压 1 006，CSN6580。

练 习 题

1. 口语练习

说明：请用给出的备选项进行替换练习。

（1）**C:** DAL367, squawk <u>3715</u>.

　　P: Squawking <u>3715</u>, DAL367.

　　　---------- low

　　　---------- normal

　　　---------- MAYDAY

　　　---------- C

（2）**C:** CCA1331, turn right heading 030 <u>for positioning</u>.

　　P: Right heading 030, CCA1331.

　　　---------- for spacing

　　　---------- for delay

　　　---------- for sequencing

　　　---------- for downwind

　　　---------- due traffic

　　　---------- to intercept localizer

2. 听说练习

说明：填空，听录音核对答案，并分角色进行口语练习。

（1）**P:** Dongfang Approach, CDG129.

　　C: CDG129, report_____.

　　P: Heading 135 at 4800 meters, CDG129.

　　C: CDG129, _____, _____.

　　P: Turning left heading 020, CDG129.

　　　（A moment later）

　　C: CDG129, _____, position 35 kilometers east of airfield.

（2）**P:** Dongfang Approach, CXA8109, maintaining 5400 meters, estimating HGH 25.

　　C: CXA8109, Dongfang Approach, report heading.

　　P: Heading 270, CXA8109.

　　C: CXA8109, _____, _____.

　　P: Turning right heading 300, CXA8109.

　　　（A moment later）

　　C: CXA8109, not _____, resume own navigation, _____ GYA.

　　P: _____, _____, CXA8109.

（3）**C:** CES2308, _____ 3403.

　　P: _____ 3403, CES2308.

　　C: CES2308, _____ , _____ present heading,

　　P: _____, CES2308.

（4）**C:** BAW201, fly _____ 330 for spacing, _____ to 2400 meters.

　　P: Heading 330, climbing to 2400 meters, BAW201.

　　C: BAW201, continue climb to 3600 meters on standard.

　　C: BAW201, continue climb to 3600 meters on standard.

　　C: BAW201, Wuhan Approach.

　　C: BAW201, _____, if you read, _____ IDENT.

　　C: BAW201, _____, will continue radar control.

（5）**C:** DLH702, Beijing Control.

　　C: DLH702, Beijing Control.

　　C: DLH702, _____ not _____, if you read, _____ right _____ 120, I say again, _____ right _____ 120.

　　C: DLH702, turn _____, resume own navigation, direct JEA, will continue radar control.

（6）**C:** SAS995, low altitude_____, _____your_____ immediately, QNH 1006, the _____flight altitude is 650 meters.

　　P: Roger, SAS995.

3. 听力练习

说明：听录音，把管制员或驾驶员通话中错误的信息和更正（或纠正）后的正确信息填写到表 3-1 中，例如：

类型 1　**C:** CHH7376, climb to 7800 meters, correction, climb to 8400 meters.

错误信息：7800 meters。更正后的正确信息：8400 meters。

类型 2　**C:** AAL304, turn right heading 120.

　　　　　P: Left heading 210, AAL304.

　　　　　C: Negative, AAL304 turn right heading 120.

　　　　　P: Right heading 120, AAL304.

错误信息：left heading 210。纠正后的正确信息：right heading 120。

表 3-1

通 话 序 号	错 误 信 息	更正（或纠正）后的正确信息
（1）		
（2）		
（3）		
（4）		

4. 听力练习

说明：听录音，在图 3-1 中标出各航空器的呼号。

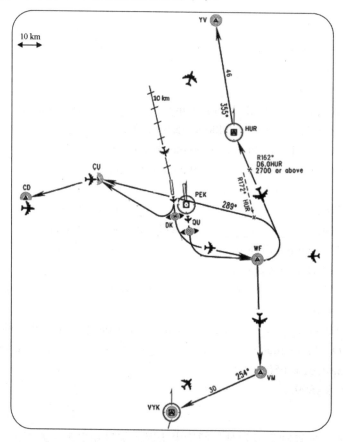

图 3-1

5. 口语练习

说明：根据图 3-2 中所给信息，请利用转弯识别法编写对话，识别 CES2328 和 CCA1325。

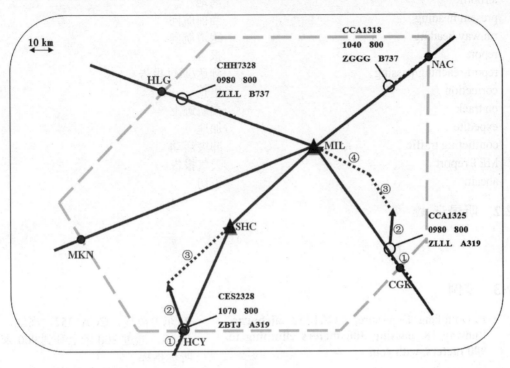

图 3-2

提示信息如下所示。

ATS unit：Dongfang Control

CES2328（从 HCY 进区域，沿航线飞往 NAC）	CCA1325（从 CGK 进区域，沿航线飞往 HLG）
① 询问当前航向、高度。	① 询问当前航向、高度。
② 在当前航向上左转 45 度。	② 在当前航向上右转 45 度。
③ 识别完成后通知驾驶员，并立刻指挥航空器恢复自主领航，飞往 SHC [位置以相对于 SHC 的方位（southwest）和距离（30 km）已给出]。	③ 识别完成后通知驾驶员，并指挥航空器左转与原航线平行飞行。
	④ 之后恢复自主领航，飞往 MIL[位置以磁航迹（300°）和距离（20 km）已给出]。

3.2　离场指令

一般来说，空中交通监视服务中航空器的离场方式主要由驾驶员自主领航离场（标准仪表离场程序和 RNAV 离场程序）、雷达引导离场和组合离场三种组成。管制员根据需要，向离场航空器发布相应的指令。

3.2.1 单词与词组

airborne	离地
present heading	当前航向
runway heading	跑道航向
report	报告
report reaching/passing	到达/通过报告
correction	更正
on track	在航迹上
expedite	加速
conflicting traffic	冲突活动
MET report	天气报告
abeam	正切

3.2.2 听录音 🔊

3.2.3 范例

（1） **P:** Dongfang Departure, CCA1352, **airborne**, Runway 18, **passing 300 meters climbing to 900 meters**, with you.

C: CCA1352, Dongfang Departure, **radar contact, follow BAV 01 Departure**, continue climb to 2700 meters on QNH 998.

P: BAV 01 Departure, climbing to 2700 meters on QNH 998, CCA1352.

（A moment later）

C: CCA1352, continue climb to 5400 meters on standard.

P: Continue climb to 5400 meters, CCA1352.

（A moment later）

P: CCA1352, 5400 meters maintaining.

C: CCA1352, contact Dongfang Control on 120.1, good day.

P: 120.1 for Control, CCA1352, good day.

（2） **P:** Dongfang Departure, CDG1835, heavy.

C: CDG1835, Dongfang Departure, identified, **cancel SID, continue runway heading**, climb to 3600 meters on standard, **after passing 3000 meters, proceed direct to P40.**

P: Continue runway heading 040, climb to 3600 meters on standard, after passing 3000 meters,

P: 东方离场，CCA1352，离地，跑道 18，通过 300 米上升到 900 米，听你指挥了。

C: CCA1352，东方离场，雷达看到，沿包头 01 号离场，继续上升到修正海压 2 700 米，修正海压 998。

P: 包头 01 号离场，上升到修正海压 2 700 米，修正海压 998，CCA1352。
（稍后）

C: CCA1352，继续上升到标准气压 5 400 米。

P: 继续上升到 5 400 米，CCA1352。
（稍后）

P: CCA1352，5 400 米保持。

C: CCA1352，联系东方区域 120.1，再见。

P: 区域 120.1，CCA1352，再见。

P: 东方离场，CDG1835，重型。

C: CDG1835，东方离场，已识别，取消标准离场，保持跑道航向，上升到标准气压 3 600 米，通过 3 000 米以后，直飞 P40。

P: 保持跑道航向 040，上升到标准气压 3 600 米，通过 3 000 米以后，

direct P40, CDG1835.

C: CDG1835, report passing 3000 meters.

P: Wilco, CDG1835.

（A moment later）

P: CDG1835, passing 3000 meters, estimating P40 1324.

C: CDG1835, roger.

C: CDG1835, contact Dongfang Control 127.3, good day.

P: 127.3, CDG1835.

（3） P: Dongfang Departure, DAL185, airborne, Runway 23R, with you.

C: DAL185, Dongfang Departure, radar contact, follow VYK 01 departure, climb and maintain 1500 meters on QNH 1003.

P: VYK 01 Departure, climbing to 1500 meters, QNH 1003, DAL185.

C: DAL185, cancel SID, turn right heading 050 due traffic.

P: Cancel SID, turn right heading 050, DAL185.

P: Dongfang Departure, DAL185, on heading 050.

C: DAL185, roger, climb and maintain 6600 meters on standard.

P: Climbing to 6600 meters on standard, DAL185.

C: DAL185, turn left direct to RENOB, resume own navigation, **magnetic track 320, distance 15 kilometers**.

P: Turn left direct RENOB, resume own navigation, DAL185.

C: DAL185, contact Dongfang Control on 125.65.

P: Contact Dongfang Control on 125.65, DAL185.

（4） P: Beijing Departure, HDA396, airborne, Runway 36L, climbing to 900 meters.

C: HDA396, Beijing Departure, **confirm squawking** 4523.

P: Affirm, HDA396.

C: HDA396, radar contact, **follow CHEDY ONE RNAV Departure**.

P: CHEDY ONE RNAV Departure, HDA396.

C: HDA396, contact Beijing Control on 128.5.

直飞 P40， CDG1835。

C: CDG1835，通过 3 000 米报告。

P: 照办，CDG1835。

（稍后）

P: CDG1835，通过 3 000 米，预计 P40 1324。

C: CDG1835，收到。

C: CDG1835，联系东方区域 127.3，再见。

P: 127.3，CDG1835。

P: 东方离场，DAL185，离地，跑道 23 右，听你指挥了。

C: DAL185，东方离场，雷达看到，大王庄 01 号离场，上升到修正海压 1 500 米保持，修正海压 1 003。

P: 大王庄 01 号离场，上升到修正海压 1 500 米，修正海压 1 003，DAL185。

C: DAL185，取消标准离场，右转航向 050，由于冲突。

P: 取消标准离场，右转航向 050，DAL185。

P: 东方离场，DAL185，航向 050。

C: DAL185，收到。上升到标准气压 6 600 米保持。

P: 上升到标准气压 6 600 米，DAL185。

C: DAL185，左转直飞 RENOB，恢复自主领航，磁航迹 320，距离 15 千米。

P: 左转直飞 RENOB，恢复自主领航，DAL185。

C: DAL185，联系东方区域 125.65。

P: 联系东方区域 125.65，DAL185。

P: 北京离场，HDA396，离地，跑道 36 左，上升到 900 米。

C: HDA396，北京离场，证实应答机 4523。

P: 是的，HDA396。

C: HDA396，雷达看到，沿 CHEDY 1 RNAV 离场。

P: CHEDY 1 RNAV 离场，HDA396。

C: HDA396，联系北京区域 128.5。

P: Control on 128. 5, HDA396.

P: 区域 128.5，HDA396。

（5）　C: CPA589, cancel LADIX altitude restriction，maintain 2700 meters due traffic.

C: CPA589，由于冲突，取消 LADIX 高度限制，保持 2 700 米。

P: Cancel LADIX altitude restriction, maintaining 2700 meters, CPA589.

P: 取消 LADIX 高度限制，保持 2 700 米，CPA589。

（6）　C: DLH457, **cleared direct AA011**, **rejoin** LADIX ONE RNAV Departure **at AA011**.

C: DLH457，可以直飞 AA011，在 AA011 重新加入 LADIX 1 RNAV 离场。

P: Direct AA011, rejoin LADIX ONE RNAV Departure, DLH457.

P: 直飞 AA011，重新加入 LADIX 1 RNAV 离场，DLH457。

（7）　C1: B3475, you are now clear of controlled airspace. Contact Dongfang Information on 129.9.
P: Dongfang on 129.9, B3475.
（A moment later）
P: Dongfang Information, B3475, MA60, departed from Dongfang Airport 1400, direct to Taiyuan Airport, VMC, climbing from 1800 meters to 3000 meters. Request any known conflicting traffic information.
C2: B3475, Dongfang Information, no reported traffic.

C1: B3475，正在离开管制空域，联系东方飞服 129.9。
P: 东方 129.9，B3475。
（稍后）
P: 东方飞服，B3475，新舟 60，1400 分从东方机场起飞，直飞太原机场，目视气象条件，从 1 800 米上升到 3 000 米，请求飞行活动情报。
C2: B3475，东方飞服，没有活动报告。

练　习　题

1. 口语练习

说明：角色扮演，两人一组。两人分别扮演管制员和驾驶员，参照样例进行通话练习。然后交换角色再次练习。例如：

> P: Dongfang Approach, CXA8102, airborne at 25, passing 130 meters climbing to 900 meters, good morning.
>
> C: CXA8102, Dongfang Approach, radar contact, via IDU-9X Departure, climb to 4500 meters on standard, expedite climb to 3000 meters.
>
> P: IDU-9X Departure, climbing to 4500 meters, expedite, CXA8102.
> （A moment later）
>
> P: CXA8102, maintaining 4500 meters.
>
> C: CXA8102, contact Dongfang Control on 120.1, good day.

P: 120.1 for Control, <u>CXA8102</u>, good day.

从表 3-2 中选取航空器呼号、离场程序代号和指令高度进行替换后练习。

表 3-2

航空器呼号	离场程序代号	指 令 高 度
ANA132	MEP-9X	5100
AZA971	OPU-08D	4500
ACA083	FQG-1D	5700
AIC372	PIK-2X	5400
CSN2356	JTG-21D	4800

2. 听说练习

说明：听录音，并分角色复述通话内容，共三段对话。

3. 听力练习

说明：听录音，补全管制员指令。

（1）**P:** Dongfang Approach, CCA4132, airborne at 56, good morning.

　　C: _____.

　　P: Climbing to 2100 meters on QNH 1011, CCA4132.

（2）**P:** Dongfang Departure, UAL971, good afternoon.

　　C: _____.

　　P: Climbing to 3600 meters on present heading, UAL971.

（3）**P:** Beijing Departure, ACA083, good morning.

　　C: _____.

　　P: Affirm, ACA083.

　　C: _____.

　　P: Left heading 030, climbing to 2700 meters, QNH 1011, ACA083.

（4）**P:** Shanghai Departure, JAL4372, good morning.

　　C: _____.

　　P: Affirm, JAL4372.

　　C: _____.

　　P: Climbing straight ahead to 2400 meters, JAL4372.

（5）**P:** Guangzhou Departure, CSN2356, on your frequency.

　　C: _____.

P: Climbing to 3000 meters, expedite, CSN2356.

4. 听力练习

1. 选择题

（1）In Dialog 1, in which altitude block should CXA8102 speed up the climb? （　　）

    A. 300 meters to 900 meters   B. 900 meters to 4500 meters

    C. 300 meters to 3000 meters   D. 3000 meters to 4500 meters

（2）According to Dialog 2, where is CSN3823 before resuming own navigation? （　　）

    A. over the runway   B. north of the airfield

    C. over DPG   D. south of the airfield

（3）According to Dialog 3, what is the departure route of CBJ812? （　　）

    A. Follow STL 01D Departure to STL

    B. After departure, fly heading 010 then resume own navigation to STL

    C. Not given

    D. Follow STL 01D Departure initially, then fly heading 010, next turn right heading 050, finally resume own navigation to STL

2. 根据所听到的内容，在图 3-3 上用表 3-3 中所指定的线型标画出每个航空器的离场路径。

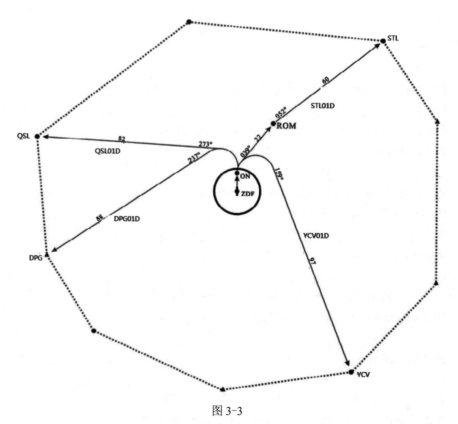

图 3-3

表 3-3

呼号	CXA8102	CSN3823	CBJ812
线型	–·– · – · –···	– –· – –· –···	– – – – – – –

5. 口语练习

说明：根据图 3-4 所示信息及以下提示，编制四段对话并口头练习（管制单位呼号：东方进近）。

（1）CES2342 起飞后告知其已被识别，指挥其按照 KG01 号离场程序飞行，并给出 3 000 米（QNH1001）的高度指令。

（2）向 CSN3443 发出 4 500 米的高度指令并要求其调到东方区域（128.35）的频率上。

（3）指挥 CSN3567 继续上升到 2 700 米，告知由于要拉开与前机的间隔，因此对其进行引导，当满足要求后（如图 3-4 中 A 位置处）给出 4 500 米的高度指令，并要求其向 ZF 归航，给出其相对于 ZF 的位置（以相对于某点距离和方位的形式）。

（4）AFR562 起飞后先指挥其沿离场航线飞行，给出 1 800 米的高度指令；之后由于与 CHH7648 的冲突，对其进行雷达引导，与原航线有 10 千米的侧向间隔后指挥航空器平行于原航线飞行，并给出 4 500 米的高度指令；冲突解除后（如图 3-4 中 B 位置处），要求其向 XS 归航，给出相对于 XS 的位置（以相对于某点磁航迹和距离的形式）；当航空器距边界点 10 千米时，要求其保持高度并调到东方区域（128.35）的频率上。

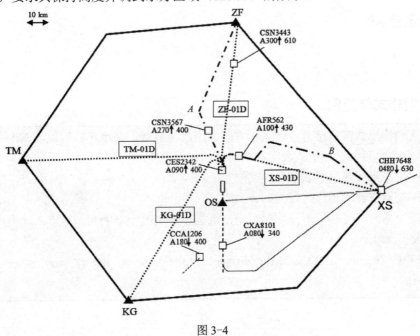

图 3-4

3.3 飞行活动通报

飞行活动通报是陆空通话中一项重要的内容。当航空器之间存在飞行冲突，或在其他一些必要的情况下，管制员通常需要给相关的航空器发送活动通报，让驾驶员保持必要的情景意识，增加更多的安全保护。

　　飞行活动通报术语与标准的管制指令略有不同，它具有简练性、灵活性、多样性等特点。其核心内容是信息，而不是操作指令，管制员通常需要根据即时的空中交通态势迅速选择通报的内容，做到精练且实用。

3.3.1　单词与词组

unknown	不明活动
opposite direction	相对飞行
same direction	顺向
11 o'clock	11 点钟方位
crossing left to right	从左向右穿越
negative contact	没有看到
vectors	引导
fast moving	快速移动
closing	接近
converging	汇聚
diverging	分散
parallel	平行
overtaking	超越

3.3.2　听录音 🔊

3.3.3　典型格式与范例

> 飞行活动情报通常包括方位、飞行方向、距离、机型、高度等，可以如下形式给出：
> （1）以 12 小时制时钟形式表示冲突活动的相对方位；
> （2）使用千米（海里）表示与冲突航空器的距离；
> （3）冲突航空器的预计飞行方向；
> （4）航空器的高度和机型（如果不明，可描述冲突航空器的相对速度，如慢或快）。
> 具体格式如下：
> TRAFFIC (number) O'CLOCK (distance) (direction of flight) [any other pertinent information].

范例：

（1）　C: AFR565, **unknown traffic, 1 o'clock, 15 kilometers, opposite direction, fast moving, report traffic in sight**.

　　P: Looking out, AFR565.
　　（A moment later）

C: AFR565，不明飞行活动，1 点钟方位，15 千米，相对飞行，快速移动，看到活动报告。

P: 正在观察，AFR565。
　　（稍后）

P: AFR565, traffic in sight.

P: AFR565，看到活动。

（2）　**C:** CSC215, unknown traffic, 10 o'clock, 11 miles, **crossing left to right**, fast moving, report traffic in sight.

C: CSC215，不明飞行活动，10 点钟方位，11 海里，从左向右穿越，快速移动，看到活动报告。

P: CSC215, negative contact, request vectors.

P: CSC215，没有看到，请求引导。

C: CSC215, turn left heading 050.

C: CSC215，左转航向 050。

P: Turning left heading 050, CSC215.

P: 左转航向 050，CSC215。

C: CSC215, **clear of traffic**, resume own navigation, direct P23.

C: CSC215，冲突解除，恢复自主领航，直飞 P23。

P: Direct P23, CSC215.

P: 直飞 P23，CSC215。

（3）　**C:** CHH7873, traffic, 10 o'clock, 12 kilometers, **northbound**, MA60, **above you**.

C: CHH7873，飞行活动通报，10 点钟方位，12 千米，向北飞行，新舟 60，比你高。

P: Looking out, CHH7873.

P: 正在观察，CHH7873。

C: CHH7873, **do you want vectors**?

C: CHH7873，需要引导吗？

P: Negative vectors, traffic in sight, CHH7873.

P: 不需要引导，看到活动，CHH7873。

（4）　**C:** AMU866, traffic at 10 o'clock, 8 kilometers, **closing from left, indicating slightly below**, fast moving.

C: AMU866，飞行活动通报，10 点钟方位，8 千米，从左侧接近，比你略低，快速移动。

（5）　**C:** CSN3369, **turn right 30 degrees immediately to avoid traffic at** 11 o'clock, 5 kilometers.

C: CSN3369，立即右转 30 度避让活动，11 点钟方位，5 千米。

P: Turning right 30 degrees, CSN3369.

P: 右转 30 度，CSN3369。

C: CSN3369, clear of traffic, resume own navigation, direct NCH.

C: CSN3369，冲突解除，恢复自主领航，直飞昌北。

P: Resume own navigation, direct NCH, CSN3369.

P: 恢复自主领航，直飞昌北，CSN3369。

（6）　**C:** CPA102, Beijing Approach, turn left immediately **heading 330 to avoid unknown traffic**, 12 o'clock, 10 kilometers, fast moving.

C: CPA102，北京进近，立刻左转航向 330，避让不明飞行活动，12 点钟方位，10 千米，快速移动。

P: Turning left heading 330, CPA102.

P: 左转航向 330，CPA102。

（7）　**C:** CSN3978, Xi'an Approach, **traffic is unmanned free balloon**, estimated over NSH **at** 19, level unknown, **moving** west.

C: CSN3978，西安进近，无人驾驶自由气球，预计过宁陕 19 分，高度不明，向西方移动。

P: Roger, traffic in sight, CSN3978.

P: 收到，看到活动，CSN3978。

练 习 题

1. 口语练习

说明：角色扮演，两人一组。两人分别扮演管制员和驾驶员，参照通话样例，运用表 3-4

中的信息进行替换后再练习。例如：

> **C:** SIA805, unknown traffic, <u>10 o'clock</u>, 10 kilometers, <u>crossing left to right</u>.
>
> **P:** Looking out, SIA805. Request vectors.
>
> **C:** SIA805, turn <u>right</u> heading <u>110</u>.
>
> **P:** <u>Right</u> heading <u>110</u>, SIA805.
>
> **C:** SIA805, now clear of traffic. Resume own navigation to <u>JCS</u>.

表 3-4

方　位	运　动　方　向	航　向	导　航　台
12 o'clock	opposite direction	right/320	PLT
1 o'clock	crossing right to left, fasting moving	left/270	BTO
2 o'clock	same direction	left/270	QJG
9 o'clock	overtaking you on your left	left/090	OD
11 o'clock	converging	right/150	GT
9 o'clock	parallel with you	left/090	YV

2. 听力练习

说明：听录音，填空。

（1）**C:** BAW250, Beijing Control, _____traffic, _____, 12 kilometers, _____
_____, Boeing 747, _____, fasting moving, _____.

　　　P: Looking out, BAW250.

　　　（A moment later）

　　　P: Beijing Control, BAW250 , _____,　request _____.

　　　C: BAW250, turn _____heading _____.

　　　P: _____, BAW250.

　　　C: BAW250, _____, _____, _____.

　　　P: Roger, _____, BAW250.

（2）**C:** THA435, Dongfang Control, _____, 8 o'clock, 28 kilometers,
_____, _____.

　　　P: Looking out, THA435.

　　　（A moment later）

　　　P: Traffic in sight, _____, _____, THA435.

　　　C: THA435,_____ .

　　　P: Roger, THA435.

3. 口语练习

说明：向图 3-5 中的标出呼号的航空器通报飞行活动。

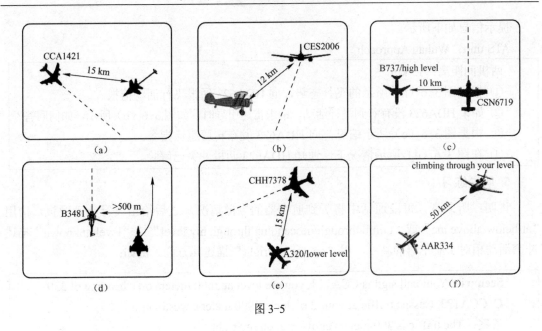

图 3-5

4. 口语练习

说明：按照图 3-6 所示的信息及所给提示，编写对话，并进行口语练习。

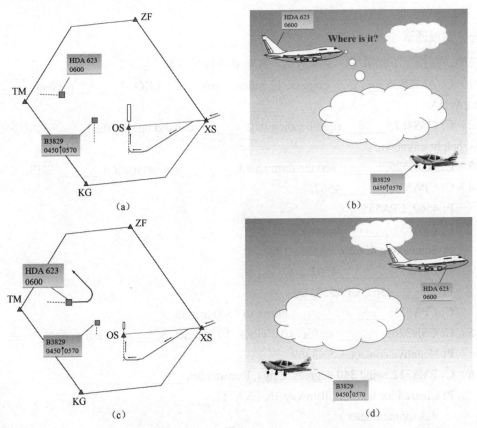

图 3-6

提示信息如下所示。

ATS unit：Wuhan Approach

> 两机间距离：15 km
> ① 根据图 3-6（a）所示的飞行态势，向 HDA623 通报飞行活动情报。
> ② 如果 HDA623 没有看到飞行活动，请求雷达引导时，如图 3-6（b）所示，如何回答？
> ③ 根据图 3-6（c）所示信息，向 HDA623 发布转弯避让指令。
> ④ 在图 3-6（d）所示情况下，通知 HDA623 两机冲突已解除。

5. 听说练习

说明：听录音，每段通话中将听到航空器的当前情况，之后给出飞行活动通报，请用 "at/below/above me" 或 "climb through/descending through my level" 或 "level unknown" 描述冲突活动相对于航空器的高度，用 "on my left/right" 描述其方位。例如：

> **Scenario**: Your call sign is CCA123, you are level at 3600 meters on a heading of 330.
> **C**: CCA123, Cessna traffic at your 2 o'clock, 3900 meters, westbound.
> 答案：The traffic is 300 meters above me, on my right.

本章综合练习

一、完成通话

说明：根据上下文填空，完成通话，并听录音核对答案。

（1）**C:** AMU525, _____ contact, 12 miles west of LKO, _____ present _____ until crossing WHA.

（2）**C:** CSS467, _____ right heading 070 _____ to avoid unidentified _____, 10 o'clock, 8 miles, height unknown.

（3）**C:** HDA525, _____ service terminated, _____ own navigation _____ SHR.

（4）**C:** CPA555, _____ 4562.

　　P: 4562, CPA555.

　　C: CPA555, _____ squawk.

　　P: Squawking 4562, CPA555.

　　C: CPA555, _____ 4562.

　　P: _____ 4562, CPA555.

　　C: CPA555, _____, climb to 4500 meters.

（5）**C:** CSN3369, _____ traffic, 10 o'clock, 10 kilometers, _____ left to right.

　　P: Negative contact, CSN3369. _____.

（6）**C:** EVA712, wind 340 degrees, 2 m/s, Runway 36, _____.

　　P: Cleared for take-off, Runway 36, EVA712.

　　　　（A moment later）

　　P: Departure, EVA712, airborne at 15, passing 100 meters climbing to 900 meters.

C: EVA712, _____, _____ to 3000 meters, report _____ 1800 meters. _____ SID, _____ 330.

P: Climbing to 3000 meters, left heading 330, wilco, EVA712.

（A moment later）

P: EVA712, passing 1800 meters.

C: _____, roger, report reaching 3000 meters.

P: Wilco, EVA712.

（A moment later）

P: EVA712, approaching 3000 meters.

C: EVA712, _____ direct JTG, _____ 60 kilometers northwest of the field.

P: Direct JTG, EVA712.

C: EVA712, climb to 5400 meters on standard, _____ 2000 ft/min _____.

P: Expediting climb to 5400 meters, EVA712.

（A moment later）

P: EVA712, maintaining 5400 meters.

C: EVA712, _____ Dongfang Control on 120.1, good day.

P: Dongfang Control on 120.1, good day, EVA712.

二、听力练习

说明：听录音，并写出所听到的指令。

（1）_____.

（2）_____.

（3）_____.

（4）_____.

（5）_____.

（6）_____.

（7）_____.

（8）_____.

（9）_____.

（10）_____.

（11）_____.

（12）_____.

（13）_____.

（14）_____.

（15）_____.

三、听力练习

1. 选择题

（1）According to Dialog 1, UAL058 is at _____ stage of flight. （   ）

　　A. cruise　　　　　　　　　　B. landing

    C. approach   D. departure

（2）According to Dialog 2, what is the reason for MGL101 making a right turn? （　　）

    A. due traffic   B. for downwind

    C. for spacing   D. for final

（3）What action does the controller take when there is a collision hazard in Dialog 4? （　　）

    A. speed adjustment

    B. instruct the aircraft to make a left turn

    C. change the level of the aircraft

    D. none of above

（4）What is the weather like at Nanping airport in Dialog 6? （　　）

    A. sunny   B. cloudy   C. not available   D. windy

2. 听录音，然后补全管制员指令。

（1）C: _____

P: Squawking 4261, UAL058.

C: _____

P: Squawking 4261, UAL058.

C: _____

P: Right heading 150, climbing to 2400 meters, LADIX One RNAV Departure, UAL058.

C: _____

P: Climbing to 5400 meters, direct AA011, LADIX One RNAV Departure, wilco, UAL058.

C: _____

P: Climbing to 4500 meters, UAL058.

（2）C: _____

P: Right heading 090, MGL101.

（3）C: _____

P: Looking out, MAS802.

P: Xi'an Approach, MAS802, traffic in sight.

C: _____

P: Roger, MAS802.

（4）C: _____

P: Negative contact, request vectors, HVN1302.

C: _____

P: Left heading 050, HVN1302.

C: _____

P: Roger, track 070, distance 27 miles, HVN1302.

（5）P: Dongfang Approach, DKH2371, passing the zone boundary.

C: _____

P: 125.75, DKH2371.

（6）**P:** Dongfang Information, G-ABCD, Yun-7, departed from Dongfang airport 1332Z, 1800 meters, VFR to Nanping airport, estimating Nanping airport 1442Z, request Nanping airport weather.

FIS: _____ .

四、故事复述

说明：听录音，并复述故事内容。

五、小拓展：鸟击

你听说过鸟击吗？鸟击的原因可能是什么？它又能造成什么后果呢？你将听到一段有关鸟击的实况录音。听录音并回答以下问题。

（1）列出与鸟击相关的单词或词组。

_____

_____ .

（2）当航空器驾驶员报告机上情况时，他报告了两次，第一次给出的信息是：

_____

_____

_____ .

第二次给出的信息是：

_____

_____

_____ .

（3）当航空器驾驶员报告意图时，他说：

_____

_____

_____ .

他还说：

_____

_____

_____ .

（4）当管制员向航空器驾驶员提供两种方案选择时，他说：

_____

_____

_____ .

他还说：

_____

_____

_____ .

（5）当管制员希望了解机上人数和剩余燃油时，他说：

_____

_____

_____ .

　　他还说：

_____

_____

_____ .

（6）当管制员希望能够详细询问机上情况时，他说：

_____

_____

_____ .

第4章 区域管制

区域管制管辖空间大，航空器飞行时间长，管制员使用的通话用语相对更加丰富。除了一般的高度、速度和航向指令外，还涉及与航路等待、缩小垂直间隔（RVSM）、侧向偏置（SLOP）、绕飞雷雨等相关的通话用语。

通过本章的学习，应达到以下学习目标：

> ❖ 掌握高度指令的发布；
> ❖ 掌握管制员要求驾驶员的额外位置报告，以及向航空器提供位置信息的方式和内容；
> ❖ 掌握等待指令的发布；
> ❖ 理解 RVSM 运行的指令；
> ❖ 理解驾驶员位置报告的内容；
> ❖ 了解绕飞雷雨的常用指令；
> ❖ 了解航空器加入、穿越或离开航路的指令。

4.1 高度指令

离场（或进近）管制员在航空器达到一定高度后，将其移交给区域管制员。区域管制员指挥航空器继续上升至巡航高度。进入巡航阶段后，因为颠簸、结冰等原因，航空器可能会申请调整高度，管制员在考虑整体交通后做出决定。某些时候，由于载荷限制，航空器可能无法执行管制员的高度指令。当航空器接近目的地时，航空器需要下降高度以实施进近着陆。

需要注意的是，在不同的飞行阶段及高度表设定基准下，高度可以是海拔高度（altitude）、高（height）或者飞行高度层（flight level）。

4.1.1 单词与词组

report	报告
reach	到达
available	可用
unavailable	不可用
further climb	继续上升
heavy traffic	交通繁忙
congested	堵塞的，拥挤的
leave	离开
rate of descent	下降率
rate of climb	上升率

maintain	保持
odd levels	单数高度层
even levels	双数高度层

4.1.2 听录音 🔊

4.1.3 范例

（1） **C:** AAL138, **report** level.
P: Maintaining 3000 meters, AAL138.

C: AAL138，报告高度。
P: 保持 3 000 米，AAL138。

（2） **C:** AIC349, **report leaving/reaching/passing** 7500 meters.
P: Wilco, AIC349.
（A moment later）
P: AIC349, leaving/reaching/passing 7500 meters.

C: AIC349，离开/到达/通过 7 500 米报告。
P: 照办，AIC349。
（稍后）
P: AIC349，离开/到达/通过 7 500 米。

（3） **C:** CBJ5154, maintain 8100 meters **until further advised**.
P: Maintaining 8100 meters, CBJ5154.

C: CBJ5154，在进一步通知前，保持 8 100 米。
P: 保持 8 100 米，CBJ5154。

（4） **C:** AAR3213, climb to 7500 meters.
P: Leaving 6300 meters, climbing to 7500 meters, AAR3213.

C: AAR3213，上升到 7 500 米。
P: 离开 6 300 米，上升到 7 500 米，AAR3213。

（5） **C:** CCA1709, Dongfang Approach, after passing RENOB, descend to 4500 meters.
P: After RENOB, descend to 4500 meters, CCA1709.

C: CCA1709，东方进近，通过 RENOB 后，下降到 4 500 米。
P: RENOB 后，下降到 4 500 米，CCA1709。

（6） **C:** CXA8586, **stop descent at** 5100 meters.
P: Stopping descent at 5100 meters, CXA8586.

C: CXA8586，在 5 100 米停止下降。
P: 5 100 米停止下降，CXA8586。

（7） **C:** CES5173, **continue climb to** 10100 meters.
P: Climbing to 10100 meters, CES5173.

C: CES5173，继续上升到 10 100 米。
P: 上升到 10 100 米，CES5173。

（8） **C:** DLH7323, recleared 9500 meters.
P: Recleared 9500 meters, DLH7323.

C: DLH7323，重新许可到 9 500 米。
P: 重新许可到 9 500 米，DLH7323。

（9） **C:** HDA901, climb to 10400 meters, **expedite**

C: HDA901，上升到 10 400 米，尽快通

until passing 9200 meters.
P: Climbing to 10400 meters, expediting until passing 9200 meters, HDA901.
（or）
P: Unable to expedite due weight, HDA901.

过 9 200 米。
P: 上升到 10 400 米,尽快通过 9 200 米,HDA901。
（或）
P: 由于超重,不能尽快,HDA901。

（10） C: CSN3201, Beijing Control, **climb at 10 m/s.**

C: CSN3201,北京区域,上升率 10 米每秒。

（11） C: CES2020, Beijing Control, climb at 2000 ft/min **or greater.**

C: CES2020,北京区域,上升率不小于 2 000 英尺分钟。

（12） C: AHK8783, Shanghai Control, climb at 1500 ft/min **or less.**

C: AHK8783,上海区域,上升率不大于 1 500 英尺分钟。

（13） C: CKK206, Beijing Control, **climb to reach** 9500 meters **by** VYK.

C: CKK206,北京区域,在大王庄前上升到 9 500 米。

（14） C: GCR6529, Beijing Control, **descend to reach** 7200 meters **at** 05.

C: GCR6529,北京区域,在 05 分下降到 7 200 米。

（15） C: CHH7988, Beijing Control, **advise if able to cross** HUR at 6300 meters.
P: Affirm, CHH7988.
C: CHH7988, **climb to and maintain** 6300 meters.

C: CHH7988,北京区域,如果能在 6 300 米通过怀柔,请通知我。
P: 是的,CHH7988。
C: CHH7988,上升到 6 300 米保持。

（16） C: CSC8825, Beijing Control, **when ready, climb to** 10100 meters.

C: CSC8825,北京区域,准备好后,上升到 10 100 米。

（17） C: ANA917, can you reach 10100 meters by VYK?

C: ANA917,能在大王庄之前到达 10 100 米吗?

（18） C: CAL503, 9500 meters is **not available** due traffic. **Can you accept** 11300 meters?

C: CAL503,由于冲突,9 500 米不可用。你能接受 11 300 米吗?

（19） C: OKA2927, **I will call you for further climb.**

C: OKA2927,进一步上升等待通知。

（20） P: CPA6111, TCAS RA.
C: CPA6111, roger, **report returning to clearance.**
（A moment later）
P: CPA6111, clear of conflict, returning to clearance. Now maintaining 7500 meters.
C: CPA6111, roger.

P: CPA6111,TCAS RA。
C: CPA6111,收到,返回许可高度报告。
（稍后）
P: CPA6111,冲突解除,正在返回许可高度,保持 7 500 米。
C: CPA6111,收到。

练 习 题

1. 词汇练习

说明：利用下框中给出的单词，完成下列句子。

| from on over away at right back to through |

（1）The crew descended _____ 3700 feet _____ reach 3000 feet.

（2）The aircraft remained _____ 6000 feet.

（3）The B747 descended _____ FL 290 to FL 210.

（4）The crew made a 180 degrees turn and flew _____ to base.

（5）ATC must maintain separation to keep aircraft _____ from each other.

（6）They stayed _____ a heading of 310 degrees.

（7）The A330 passed _____ the outer maker at 22:17.

（8）The A321 turned _____ .

2. 口语练习

说明：结对活动。首先，学生 A 读出 Student A（本书 186～187 页）的内容，学生 B 听并完成下列（1）～（5）句。然后两人交换角色，由学生 B 读出 Student B（本书 186 页）的内容，学生 A 听并完成下列（6）～（10）句。练习完成后，请听录音并跟读。

（1）ALK866, _____ to _____ meters at _____ ft/min or _____ .

（2）UAL ____ , turn _____ heading 120, descend to _____ meters.

（3）CCA891, descend to _____ meters, contact _____ on _____ .

（4）AFL209, _____ ANRAT, descend to _____ meters.

（5）ABW _____ , descend to _____ meters, expedite until passing _____ meters.

（6）AZA98, climb to _____ meters, expedite until passing _____ meters.

（7）AIC356, climb to reach _____ meters by VYK, report _____ .

（8）HVN_____ , turn _____ heading 130, descend to and maintain _____ meters.

（9）AFL65, the preceding traffic reported _____ at 9200 meters. Can you accept _____ meters?

（10）JAL _____ , descend to reach _____ meters at _____ .

3. 完成通话

说明：根据提示完成通话。

（管制员向 CCA1331 发指令，上升到 8 900 米，并通知其在到达 8 900 米后报告。）

C: _____ .

P: Climbing to 8900 meters, CCA1331.

P: CCA1331, approaching 8900 meters.

（管制员向 CCA1331 发指令，继续上升到 10 100 米。）

C: _____ .

P: Climbing to 10100 meters. Is 11300 meters available? CCA1331.

（管制员告知 CCA1331 高度层 11 300 米此时不可用，并提供 10 700 米给 CCA1331 选择。）

C: _____?

P: Negative.

（管制员向 CCA1331 发指令，通信移交至下一管制单位：上海区域，频率 124.1。）

C: _____.

P: 124.1 for Shanghai Control, CCA1331.

4. 听力练习

说明：听录音，把管制员的指令补充完整。

（1）**C1:** _____.

　　P: 120.1, CCA1527, good bye.

　　（A moment later）

　　P: Dongfang Control, CCA1527, good morning.

　　C2: _____.

　　P: Climbing to 4500 meters, heading 260, CCA1527.

　　C2: _____.

　　P: Climbing to 6300 meters, maintaining present heading, CCA1527.

　　（A moment later）

　　P: CCA1527, reaching 6300 meters.

　　C2: _____.

　　P: Climbing to 7500 meters, CCA1527.

　　C2: _____.

　　P: 129.4, CCA1527, good day.

（2）**C:** _____.

　　P: Squawking IDENT, AFR116.

　　C: _____.

　　P: Leaving 9800 meters descending to 7800 meters, AFR116.

　　C: _____.

　　P: Descending to 6000 meters, wilco, AFR116.

　　P: AFR116, passing 6600 meters, descending.

　　C: _____.

　　P: Affirm, AFR116.

（3）**C1:** _____.

　　P: 126.5, good day, ACA018.

　　（A moment later）

　　P: Beijing Control, ACA018, good morning.

　　C2: _____.

　　P: ACA018, 4800 meters maintaining, heading 230.

　　C2: _____.

P: Climbing to 9200 meters, expedite until passing 7200 meters, ACA018.

C2: _____ .

P: Left heading 180, ACA018.

C2: _____ .

P: 128.9, good day, ACA018.

（4）**C:** _____ .

P: Climbing to 8100 meters, UAL889. Is 8900 meters available?

C: _____ ?

P: Negative, UAL889.

C: _____ .

P: Climbing to 8100 meters, UAL889.

（A moment later）

P: UAL889, reaching 8100 meters.

C: _____ .

P: Dongfang Control on 128.9, good day, UAL889.

（5）**P:** Dongfang Control, CES2020, over TER, maintaining 9800 meters, with you.

C: _____ .

P: Leaving 9800 meters descend to 8400 meters, CES2020.

C: _____ .

P: Descending to 5400 meters, CES2020.

（A moment later）

P: CES2020, passing 7200 meters.

C: _____ .

P: Affirm, CES2020.

4.2　位置报告与位置信息

　　航空器飞越强制性报告点时，按照规定，需要进行位置报告，除非管制员要求其省略位置报告（雷达管制时）。管制员如果觉得有必要，也可以要求航空器进行额外的位置报告。无论如何，在航空器离开管制范围之前，都需要让其恢复正常的位置报告。

　　雷达管制环境下，管制员有时需要为航空器提供位置信息，以帮助驾驶员准确掌握自己的位置。

4.2.1　单词与词组

omit position reports	省略位置报告
intercept	截获
pass	通过
radial	径向线
resume position reporting	恢复位置报告
over	在……上空；过……

abeam　　　　　　　　　　　　正切

4.2.2 听录音 🔊

4.2.3 典型格式及范例

1. 驾驶员进行位置报告的格式

（1）航空器呼号；
（2）报告点名称；
（3）过报告点时间；
（4）当前高度；
（5）下一报告点名称；
（6）预计过报告点时间；
（7）再下一个报告点名称。

范例 **1**：

（1）**P:** Beijing Control, CCA1212, **over YQG, at** 15, maintaining 8400 meters, **estimating** BTO at 35, next VYK .

P: 北京区域，CCA1212，遥墙上空，15 分，保持 8 400 米，预计泊头 35 分，下一个点大王庄。

（2）**P:** Beijing Control, CSN6723, over BTO, at 35, maintaining 8400 meters, estimating VYK at 50.

P: 北京区域，CSN6723，泊头上空，35 分，保持 8 400 米，预计大王庄 50 分。

（3）**C:** CXA866, Wuhan Control, next report at LKO.

C: CXA866，武汉区域，下一次在龙口报告。

（4）**C:** CKK210, **omit position reports until over** JR.

C: CKK210，省略位置报告直到过良乡。

（5）**C:** CAL503, **resume position reporting**.

C: CAL503，恢复位置报告。

（6）**C:** CHH7896, report passing WXI.

C: CHH7896，过魏县报告。

（7）**C:** GCR6500, report 30 kilometers from HYN DME.

C: GCR6500，距花垣 DME 30 千米报告。

（8）**C:** ANA917, report passing 320 radial LJG VOR.

C: ANA917，通过连江 VOR 320 径向线报告。

（9）**C:** CES5321, report distance from NJL DME.
P: 40 kilometers from NJL DME, CES5321.

C: CES5321，报告距禄口 DME 的距离。
P: 距禄口 DME 40 千米，CES5321。

2. 管制员向驾驶员通报航空器位置的方式

（1）相对于一个大众所知的地理位置；

（2）相对于一个重要点、航路导航台或进近设施的磁航迹与距离；

（3）相对一个大众所知位置的方位与距离；

（4）最后进近时相对于接地端的距离；

（5）相对于 ATS 航路中心线的距离与方位。

范例 2：

（1）**C:** CCA1901, **position over** HO.　　　　C: CCA1901，位置在长武上空。

（2）**C:** BAW297, resume own navigation, direct DKO, **magnetic track** 073, **distance** 37 kilometers.　　　C: BAW297，恢复自主领航，直飞磴口，磁航迹 073，距离 37 千米。

（3）**C:** CXA8459, position 25 kilometers southwest of the field.　　　C: CXA8459，位置距机场西南 25 千米。

（4）**C:** CAL2811, **position** 20 kilometers **from touchdown**.　　　C: CAL2811，位置距接地点 20 千米。

（5）**C:** AMU127, position 10 kilometers left of the route.　　　C: AMU127，位置距航路左侧 10 千米。

练 习 题

1. 听力练习

说明：听录音，根据驾驶员的位置报告，在图 4-1 中将各航空器经过的点连线。

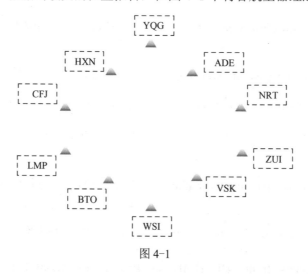

图 4-1

2. 完成通话

说明：根据中文提示，写出管制员正确的英文指令。

（1）通过 VICIS 点报告。

_____.

（2）在 DELTA 点前省略位置报告。

_____.

（3）通过 ROM VOR 210 径向线报告。

_____.

（4）距 MKE DME 15 海里报告。

_____.

（5）恢复位置报告。

_____.

（6）位置距接地点 18 千米。

_____.

（7）位置在机场东北 38 千米。

_____.

（8）位置在 A593 以东 35 千米。

_____.

（9）磁航迹 030，距离 12 千米。

_____.

（10）CSN1232，现在正通过机场上空。

_____.

3. 听力练习

说明：听录音，补全管制员指令。

（1）C: _____
   P: ECH at 35, maintaining 9200 meters, estimating WISKY at 53, CES5115.
   C: _____

（2）C: _____
   P: Wilco, HDA306.

（3）C: _____
   P: Roger, PAL331.
   （A moment later）
   C: _____
   P: Roger, PAL331.

（4）C: _____
   P: Roger, ACA085.

（5）C: _____
   P: 47 kilometers from ALF VOR, AFR129.

（6）**C:** _____.

　　　P: Roger, BAW716.

4. 完成通话

说明：按照标准陆空通话将下列短语组成驾驶员位置报告的正确格式。

（1）**P:** N-B241E, Dongfang Control, 9500 meters, at 50, over G, ETO, at 0215, H.

_____

（2）**P:** Dongfang Control, ZK-ELC, at 01, over O, 6000 meters, estimating, at 55, P.

_____

（3）**P:** D-RSTU, Dongfang Control, over W, 6600 meters, at 25, ETO, X, at 1720.

_____

4.3　航路等待

　　由于天气变化、飞行冲突或非常规情况等的发生，管制员可能需要指挥航空器在航路上等待，加入标准等待程序。某些情况下，管制员也可以发布 orbit left/right 等指令，以 360 度转弯的机动方式延长航空器的运行时间。当管制员发送等待指令时，一般应告知原因。

4.3.1　单词与词组

hold	等待
right hand circuit (pattern)	右航线
inbound	向台
outbound	背台
deteriorate	恶化
thunderstorm	雷暴
congestion	交通拥堵
VIP flight	要客
orbit	盘旋
hold visual over (position)	在（位置）上空目视等待

4.3.2　听录音 📢

4.3.3　范例

（1）　**C:** ACA087, **cleared to** WTM, descend to 8100 meters, **hold as published**, expect approach clearance at 13.

　　　C: ACA087，可以飞往 WTM 等待，下降到 8 100 米，加入标准等待程序，预计进近许可 13 分。

P: Descending to 8100 meters, hold at WTM, ACA087.

P: 下降到 8 100 米，在 WTM 等待，ACA087。

(2) **P:** Dongfang Control, BAW038, request detailed holding instructions.

C: BAW038, **proceed to** TOL, maintain 7200 meters, **hold inbound track** 038 degrees, **left hand pattern**, **outbound time** one and a half minutes, **expect further clearance at** 19.

P: 东方区域，BAW038，请求详细等待指令。
C: BAW038，可以飞往 TOL 等待，保持 7 200 米，入航航迹 038 度，左航线，出航时间一分半钟，预计进一步许可 19 分^①。

(3) **P:** Dongfang Control, AFR182, request detailed holding instructions.

C: AFR182, cleared to the 120 radial of the SGM VOR at 30 kilometers DME Fix. Maintain 6900 meters, **hold east, right hand circuit**, outbound time one and a half minutes, expect further clearance at 36, the holding speed at or below 550 km/h.

P: 东方区域，AFR182，请求详细等待指令。
C: AFR182，可以飞往 SGM VOR 120 度径向线距台 30 千米处等待。保持 6 900 米，在东面等待，右航线，出航时间一分半钟，预计进一步许可 36 分，等待速度不大于 550 千米每小时。

(4) **P:** Dongfang Control, CSN2182, request detailed holding instructions.

C: CSN2182, cleared to the 120 radial of the XSH VOR at 30 kilometers DME Fix. Maintain 6900 meters, hold between 25 and 35 kilometers DME, right hand circuit, expect approach clearance at 36.

P: 东方区域，CSN2182，请求详细等待指令。
C: CSN2182，可以飞往 XSH VOR 120 度径向线距台 30 千米处等待。保持 6 900 米，在距台 25 到 35 千米之间等待，右航线，预计进近许可 36 分。

练 习 题

1. 口语练习 1

说明：角色扮演，两人一组。两人分别扮演管制员和驾驶员，参照通话样例练习。然后交换角色再次练习。例如：

C: <u>PAL331,</u> maintain 9500 meters, hold at <u>MZL</u> as published.

P: <u>PAL331,</u> we're not familiar with the holding procedure, could you give the details?

C: <u>PAL331,</u> proceed to <u>MZL</u>, maintain 9500 meters, hold inbound track 150 degrees, left hand pattern, outbound time one and a half minutes, expect further clearance at 19.

---

① 等待指令内容的发送次序为：等待点+等待高度+入航航迹+左（右）转+出航时间。

P: Roger, <u>PAL331</u>, proceed to <u>MZL</u>, maintain 9500 meters, inbound track 150 degrees, left hand pattern, outbound time one and a half minutes.

从表 4-1 中选取航空器呼号和定位点进行替换后练习。

表 4-1

航空器呼号	定 位 点
AAR384	JNJ
AAL576	ENH
SAS995	FJC
ANA840	DSH
MAS562	NJL

2. 口语练习 2

说明：根据图 4-2 所示信息，向 AFR182 发布等待指令。等待高度为 7 000 英尺，当前时间为 10:10，预计等待时间 10 分钟。

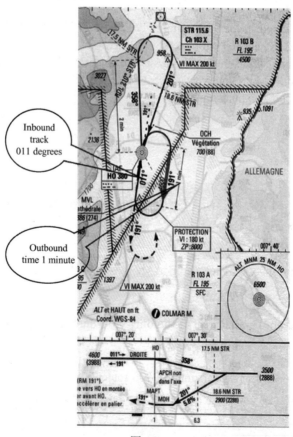

图 4-2

3. 听力练习

说明：听录音，在图 4-3 中标出管制员分别发布给 CDG4951、AAL289 和 CPA636 的详细等待航线。

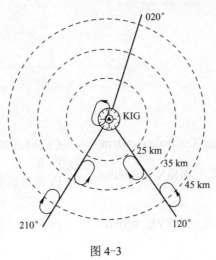

图 4-3

4.4 RVSM 运行与 SLOP

缩小垂直间隔（reduced vertical separation minimum，RVSM）是相对于常规垂直间隔而言的。RVSM 空域在我国指的是飞行高度层为 8 900 米（含）至 12 500 米（含）之间的空域。

策略横向偏置程序（SLOP）允许具备横向偏置能力且符合相关要求的航空器在 RVSM 空域内沿航路（航线）飞行时，向航路中心线（航线）右侧平行偏置一定距离，如图 4-4 所示。在 SLOP 通话术语中，"偏置（proceed offset）"是关键用语。

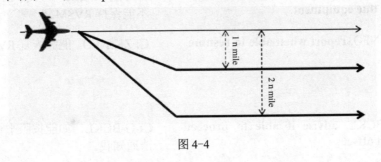

图 4-4

4.4.1 单词与词组

RVSM	缩小垂直间隔
offset	偏置
issue	发布
state aircraft	国家航空器[①]

---

① 根据《国际民航公约》规定，用于军事、海关和警察部门的航空器属于国家航空器。

turbulence	颠簸
resume	恢复
request right offset	请求向右偏置

4.4.2　听录音🔊

4.4.3　范例

1. RVSM

（1）　**C:** G-KSIX, Dongfang Control, **confirm RVSM approved**.
P: Dongfang Control, G-KSIX, negative RVSM.

C: G-KSIX，东方区域，证实 RVSM 已批准。
P: 东方区域，G-KSIX，不是 RVSM。

（2）　**P:** Dongfang Control, P-ZRVI, **affirm RVSM**.

P: 东方区域，P-ZRVI，是 RVSM。

（3）　**P:** Dongfang Control, G-FPHZ, **negative RVSM, state aircraft**.

P: 东方区域，G-FPHZ，不是 RVSM，国家航空器。

（4）　**C:** F-BCDE, **unable issue clearance into RVSM airspace**, maintain 8400 meters.

C: F-BCDE，不能进入 RVSM 空域，保持 8 400 米。

（5）　**P:** Dongfang Control, P-ZRVI, **unable RVSM due turbulence**.

P: 东方区域，P-ZRVI，由于颠簸，不能保持 RVSM。

（6）　**P:** Dongfang Control, D-BVIQ, **unable RVSM due equipment**.

P: 东方区域，D-BVIQ，由于设备原因，不能保持 RVSM。

（7）　**C:** ZK-NFO, **report when able to resume RVSM**.

C: ZK-NFO，能够恢复 RVSM 时报告。

2. SLOP

（1）　**C:** G-BCKL, **advise if able to proceed parallel offset**.

C: G-BCKL，如能按照平行偏置飞行，请通知我。

（2）　**C:** AFR116, after EPGAM, **proceed offset 2 miles right of A593**.

C: AFR116，过 EPGAM 后，向 A593 右偏置 2 海里。

（3）　**C:** ANA1260, **cancel offset**, back to A593.

C: ANA1260，取消偏置，回到 A593 航路。

（4）　**C:** CCA1061, **cancel offset**, proceed direct to GLN.

C: CCA1061，取消偏置，直飞观澜。

（5）　**C:** CSC3315, offset within 10 km left of track approved.

C: CSC3315，向左偏置 10 千米以内同意。

（6）**C:** CXA8239, right offset not approved due traffic.

C: CXA8239，由于冲突，不同意向右偏置。

练 习 题

1. 完成通话

说明：根据中文提示，写出正确的英文通话指令。

（1）当航空器将首次进入 RVSM 空域时，管制员应证实航空器是否符合 RVSM 运行，请向 CCA1324 发布如下指令：

CCA1324，证实 RVSM 已批准。

_____

（2）未获准 RVSM 运行的航空器不得在 RVSM 空域内运行，请向 N-898EF 发布如下指令：

N-898EF，不能进入 RVSM 空域，保持 8 100 米。

_____

2. 完成通话

说明：根据 4.4.3 节范例 2.SLOP 中所给出的通话例句，为 CBJ5323 编写一段完整的涉及 SLOP 的陆空通话。

4.5 绕飞雷雨

空中遇有雷雨时，在情况允许条件下，驾驶员通常采取绕飞的措施。在确认雷雨不再对航空器安全产生影响的情况下，管制员可指挥航空器回到航路。

4.5.1 单词与词组

detour	绕飞
go round	绕飞
circumnavigate	绕飞
deviate	偏离
CB (cumulonimbus)	积雨云；雷暴
build-up	积雨云；雷暴
adverse weather	恶劣天气
hazardous weather	恶劣天气
severe weather	恶劣天气
turbulence	颠簸
icing	积冰
anti-icing	防冰
de-icing	除冰
thunderstorm	雷暴

| clear air turbulence (CAT) | 晴空颠簸 |
| deteriorate | 恶化 |

4.5.2 听录音 🔊

4.5.3 范例

（1） **C:** How many miles do you need to be on present heading to circumnavigate the CB?

C: 你还需要保持当前航向多少海里来绕飞雷暴？

（2） **C:** How far do you need to track out on present heading?

C: 你还需要从当前航向出航多远距离？

（3） **C:** You can deviate 10 miles right of track to avoid build-up.

C: 可以往航迹右偏 10 海里绕飞雷暴。

（4） **C:** When clear of weather, proceed direct to BSE, resume own navigation.

C: 脱离不利天气后，直飞百色，恢复自主领航。

（5） **C:** Follow the track of the preceding aircraft to avoid build-up.

C: 跟着前机航迹绕飞雷暴。

（6） **C: Report flight conditions.**

C: 报告飞行条件。

练 习 题

1. 完成通话

说明：根据提示，写出管制员正确的英文指令。

（1） **P:** Beijing Control, CSN6721, request right turn to avoid build-up.

（管制员向 CSN6721 发指令，同意绕飞，并通知其脱离不利天气后报告。）

　　C: _____.

（2）（管制员告知 LKE9837 雷暴正快速向其方向移动，并指令其从当前位置右转 20 度飞 30 千米绕飞雷暴。）

　　C: _____.

（3） **P:** Beijing Control, GCR7891, we have an indication of hazardous weather 50 kilometers ahead of us, request clearance to go round it.

（管制员指令 GCR7891 右转绕飞雷暴。）

　　C: _____.

（4）（管制员告知 CCA1310 前机报告 8 100 米附近均有轻度颠簸。）

　　C: _____.

（5）**P:** Beijing Control, CES2831, at 10100 meters, 40 kilometers to VYK, encountering moderate icing.

（管制员告知 CES2831 如果条件继续或恶化，请随时通知管制员。）

C: _____.

2. 听力练习

说明：听录音，根据所听到的内容选择最佳答案。

（1）What is the frequency of Dongfang Control?（　　）

    A. 120.8　　　　　B. 123.5　　　　　C. 124.5　　　　　D. 126.4

（2）What is the destination level of CCA1527?（　　）

    A. 10100 meters　　　　　　　B. 6000 meters

    C. 8100 meters　　　　　　　　D.10400 meters

（3）What is the heading of CCA1331 when it requests for circumnavigation?（　　）

    A. 120　　　　　　B. 230　　　　　　C. 110　　　　　　D. 250

（4）Where is CCA1331 after 20 kilometers on the new heading?（　　）

    A. Over KG　　　　　　　　　B. 20 kilometers east of A493

    C. 15 kilometers east of A493　　D. 35 kilometers north of A593

（5）Why does CSN6903 request for higher level?（　　）

    A. Due economical reasons　　　B. Due turbulence

    C. Due traffic　　　　　　　　D. Due restrictions

4.6　航空器加入、穿越或离开航路

当航空器要加入、穿越或离开航路时，必须向管制员提出申请，只有得到管制员的许可之后才可以执行。在相关的通话术语中，"加入""穿越""离开"是关键词汇。

4.6.1　单词与词组

join　　　　　　　　　　　　加入
cross　　　　　　　　　　　穿越
leave　　　　　　　　　　　离开
rejoin　　　　　　　　　　重新加入
on track　　　　　　　　　在航迹上

4.6.2　听录音 🔊

1. 加入航路

2. 穿越航路

3. 离开航路

4.6.3 范例

（1）**C: CCA1730, join** A1 at DAL at 5400 meters.

C: CCA1730，在 DAL 加入 A1，高度 5 400 米。

（2）**C: CES5166, cleared to leave** A1 via DAL VOR.

C: CES5166，可以经由 DAL VOR 离开 A1。

（3）**C: ACA026, cleared to cross** A1 at DAL VOR at 4500 meters.

C: ACA026，可以在 DAL VOR 穿越 A1，高度 4 500 米。

练　习　题

听力练习

说明：听录音，把管制员的指令补充完整。

（1）**P:** Dongfang Control, C-GNDK, request airways crossing V22 at KOL.

　　C: _____.

　　P: C-GNDK, MA60, passing JUI heading 290, maintaining 1800 meters, VMC, ETO KOL at 1115.

　　C: _____.

　　P: Cross V22 at KOL not before 1115 at 1800 meters, C-GNDK.

（2）**P:** Dongfang Control, N550SD, request joining clearance at GOF.

　　C: _____.

　　P: N550SD, Cessna 310, 19 miles west of GOF intersection, heading 090, maintaining 2400 meters, outbound from HTE VOR/DME, ETO GOF at 45, B2 for Beian. Request 3000 meters.

　　C: _____.

本章综合练习

一、听说练习

说明：听录音，用下划线标注每句指令中强调的部分，然后听录音跟读。

（1）Climb to 5400 meters immediately.

（2）Turn left heading 230.

（3）Go around, I say again, go around. Traffic on 34 Right.

（4）Climb to 9200 meters, expedite until passing 7200 meters.

（5）Descend to 2400 meters, report passing 3600 meters.

二、听力练习

说明：收听两段等待指令，分别回答下列问题。（1）～（3）对应指令 1，（4）和（5）对应指令 2。

（1）N-3421E 的高度指令是上升还是下降？

（2）N-3421E 在 KIG 的高度是多少？

（3）N-3421E 的入航边是多少？

（4）ZK-DAE 的入航边是多少？

（5）ZK-DAE 的出航限制是多少？

三、完成通话

说明：听录音，完成管制员的通话内容。

C1: _____.

P: Line up, CPA7402.

C1: _____.

P: Cleared for take-off, Runway 18, CPA7402.

C1: _____.

P: 119.7, CPA7402.

P: Dongfang departure, CPA7402, climbing to 2100 meters, passing 1200 meters.

C2: _____.

P: Climbing to 6300 meters, CPA7402.

C2: _____.

P: 134.7, CPA7402.

P: Dongfang Control, CPA7402, climbing to 6300 meters.

C3: _____.

P: Climbing to 9500 meters, CPA7402. Any reports of turbulence at 9500 meters this morning?

C3: _____.

P: Roger, thanks, CPA7402.

四、完成通话

说明：根据上下文，把管制员的通话补充完整，然后听录音核对答案。

C0: _____ .

P: Dongfang Control 120.8, CCA1331, good day.

　（A moment later）

P: Dongfang Control, CCA1331, good morning.

C1: _____ .

P: CCA1331, over HTO, maintaining 4500 meters.

C1: _____ .

P: Climbing to 6300 meters, CCA1331.

　（A moment later）

P: CCA1331, maintaining 6300 meters.

C1: _____ .

P: Climbing to 8900 meters, expediting to 7500 meters, CCA1331.

　（A moment later）

P: CCA1331, approaching 8900 meters.

C1: CCA1331, roger.

　（A moment later）

C1: _____ .

P: Beijing Control 124.3, CCA1331, good day.

　（A moment later）

P: Beijing Control, CCA1331, good morning.

C2: _____ .

P: CCA1331, passing GOL 03, maintaining 8900 meters, estimating INK at 18.

C2: _____ .

　（A moment later）

P: CCA1331, request climb to 10100 meters due to moderate air turbulence.

C2: _____ .

P: Affirm, CCA1331.

C2: _____ .

P: Climbing to 10700 meters, CCA1331.

　（A moment later）

P: CCA1331, maintaining 10700 meters.

C2: _____ .

P: 131.1 for Shanghai, CCA1331, good day.

　（A moment later）

P: Shanghai Control, CCA1331, over MKE, maintaining 10700 meters.

C3: _____ .

　（A moment later）

C3: _____ .

P: Descending to 8400 meters, not less than 2000, CCA1331.

（A moment later）

C3: _____.

P: Descending to 6000 meters, will report reaching, CCA1331.

　（A moment later）

P: CCA1331, maintaining 6000 meters.

C3: _____.

P: 119.3 for Shanghai Arrival, CCA1331, good day.

五、听说练习

1. 听录音，回答问题。

（1）When does CES2328 fly over MIDOX?

（2）How does the controller solve the conflict between CSZ9886 and CXA8459?

（3）What does the controller say to CXA8459 when the traffic is no factor?

2. 再听一遍录音，在图4-4上用表4-2所示的不同线型标出航空器的飞行路径。

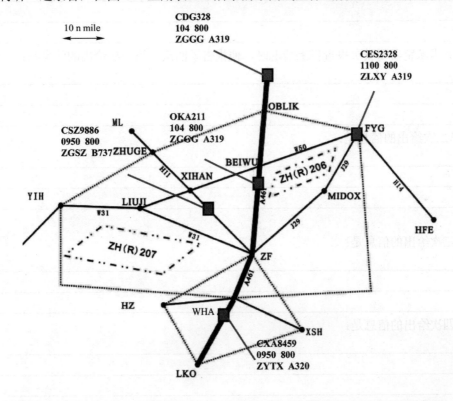

管制单位呼号：东方区域
各航班的飞行计划信息：
CES2328: FYG→J29→MIDOX→J29→ZF→A461→WHA→A461→LKO.
OKA211:OBLIK→A461→ZF→A461→WHA→A461→LKO.
CDG328: OBLIK→A461→ZF→A461→WHA→A461→LKO.
CSZ9886:ZHUGE→H11→XIHAN→H11→ZF→A461→WHA→A461→LKO.
CXA8459:LKO→A461→WHA→A461→ZF→A461→BEIWU→A461→OBLIK.

图4-4

表 4-2

呼　号	线　型
CES2328	···
CDG328	▬ · ▬ · · ▬ · ▬
CXA8459	▬ · · ▬ · ▬ · ▬ · ▬
CSZ9886	▬ ▬ ▬ ▬ ▬ ▬ ▬ ▬ ·

六、故事复述

说明：听录音，并复述故事内容。

七、小拓展：空中失火。

你听说过空中失火的情况吗？空中失火的原因可能是什么？它又能造成什么后果呢？你将听到一段有关疑似空中失火的实况录音。听录音并回答以下问题。

（1）列出与空中失火相关的单词或词组。

_____

_____.

（2）当航空器驾驶员报告机上情况时，他报告了四次，第一次给出的信息是：

_____

_____.

第二次给出的信息是：

_____

_____.

第三次给出的信息是：

_____

_____.

第四次给出的信息是：

_____

_____.

（3）当管制员希望证实机上情况时，他说：

_____

_____.

（4）当管制员询问航空器驾驶员是否需要帮助时，他说：

_____

他还说：

（5）当管制员希望了解机上人数和剩余燃油时，他说：

（6）在这起事件中，有三架其他航空器受到了影响，管制员分别对它们说：

Endeavor 3696：

Delta 405：

Delta 1：

第 5 章　进近管制——进场阶段

在进近管制进场阶段，航空器的高度、速度和航向变化频度较高，因此关于高度、速度和航向的管制指令使用较多，且通话频率较快。此外，对应不同的进场和进近方式，可使用相应的术语进行表述。

通过本章的学习，应达到以下学习目标：

> ❖ 掌握进场雷达引导的相关用语；
> ❖ 掌握航空器高度调整的相关用语；
> ❖ 掌握航空器速度调整的相关用语；
> ❖ 掌握航空器进场及进近的相关用语；
> ❖ 了解 VFR 进场的相关用语；
> ❖ 了解雷达进近的相关用语。

5.1　进场及进近

当航空器从区域移交至进近管制单位后，管制员需要根据交通情况，为航空器安排着陆次序，及时调整航空器的高度、速度和航向，若有延误，应及时发布延误指令。

5.1.1　单词与词组

ILS	仪表着陆系统
reduce speed to	减速到
resume normal speed	恢复正常速度
maintain present speed	保持现在速度
reduce to minimum approach speed	减到最小进近速度
reduce to minimum clean speed	减到最小光洁速度
no（ATC）speed restrictions	无（ATC）速度限制
vectoring for ILS approach	雷达引导 ILS 进近
vectoring for（position in the circuit）	雷达引导到（起落航线位置）
delay	延误
no delay expected	预计没有延误
holding pattern	等待航线
delay not determined	延误没有确定
straight-in approach	直线进近
terminated	终止
extend downwind	延长三边
cleared for ILS approach	可以 ILS 进近

report established 建立报告
report fully established 完全建立报告
Final 五边（进近管制单位的一个席位）

5.1.2 听录音 🔊

5.1.3 范例

（1） **C:** AFR102, **report speed**.
P: Speed 250 knots, AFR102.
C: AFR102, **reduce to minimum clean speed**.
P: Reducing to 210 knots, AFR102.

C: AFR102，报告速度。
P: 速度 250，AFR102。
C: AFR102，减到最小光洁速度。
P: 减速到 210，AFR102。

（2） **C:** HDA786, **maintain** 250 knots **or greater**, No.1 to land.
P: 250 knots, HDA786.
（A moment later）
C: HDA786, **resume normal speed**.
P: Normal speed, HDA786.

C: HDA786，保持速度不小于 250，第一个落地。
P: 250，HDA786。
（稍后）
C: HDA786，恢复正常速度。
P: 正常速度，HDA786。

（3） **C:** AFL3820, **do not exceed** 190 knots, preceding traffic is B777, heavy, **caution wake turbulence**.
P: Not exceed 190 knots, AFL3820.

C: AFL3820，速度不得超过 190，前机 B777，重型，注意尾流。
P: 速度不超过 190，AFL3820。

（4） **C:** KLM4453, Beijing Approach, **maintain present speed**.
P: Maintain present speed, KLM4453.

C: KLM4453，北京进近，保持当前速度。
P: 保持当前速度，KLM4453。

（5） **C:** CSN6136, **no speed restriction**, contact Final 118.2, good day.
P: Roger, 118.2, good day, CSN6136.

C: CSN6136，无速度限制，联系五边 118.2，再见。
P: 收到，118.2，再见，CSN6136。

（6） **C:** CES5166, **reduce to minimum app-roach speed**.
P: Reducing to minimum approach speed, CES5166.

C: CES5166，减到最小进近速度。
P: 减到最小进近速度，CES5166。

（7） **C:** CSN3102, **confirm RNAV approved**.
P: Affirm RNAV, CSN3102.
C: CSN3102, follow ATAGA1A RNAV

C: CSN3102，证实 RNAV 已批准。
P: 是 RNAV，CSN3102。
C: CSN3102，沿 ATAGA1A RNAV

Arrival.

进场。

（8）　P: Guangzhou Approach, CPA5782, **unable RNAV due equipment**.
C: CPA5782, roger, **report able to resume RNAV**.
P: Wilco, CPA5782.

P: 广州进近，CPA5782，由于设备原因不能保持 RNAV。
C: CPA5782，收到，能够恢复 RNAV 时报告。
P: 照办，CPA5782。

（9）　C: EVA707, turn left heading 090, **report runway in sight**, expect visual approach, Runway 19.
P: Roger, expect visual approach, will report when runway in sight, EVA707.
（A moment later）
P: Runway in sight, EVA707.
C: EVA707, follow the A340, **cleared for visual approach**, Runway 19, **caution wake turbulence**, report final.
P: Cleared for visual approach, Runway 19, report final, EVA707.

C: EVA707，左转航向 090，看到跑道报告，预计目视进近，跑道 19。

P: 收到，预计目视进近，看到跑道报告，EVA707。
（稍后）
P: 看到跑道，EVA707。
C: EVA707，跟在 A340 后面，可以目视进近，跑道 19，注意尾流，五边报告。
P: 可以目视进近，跑道 19，五边报告，EVA707。

（10）　C: HVN506, preceding traffic B777 at 9 o'clock, 12 kilometers, report traffic in sight.
P: **Traffic in sight**, HVN506.
C: HVN506, succeeding traffic 13 kilometers behind you, it'll cross your level by visual separation.
P: Roger, HVN506.

C: HVN506，前机是 B777，在 9 点钟方位，12 千米，看到活动报告。

P: 看到活动，HVN506。
C: HVN506，后机距离你 13 千米，它将使用目视间隔穿越你的高度层。

P: 收到，HVN506。

（11）　C: CHH7702, **turn left heading 330 immediately to avoid traffic deviating from adjacent approach**, climb to 1200 meters.

C: CHH7702，立即左转航向 330，避让偏离邻近进近航迹的飞行活动，上升到 1 200 米。

（12）　C: ANA5701, I will **take you through the localizer** for separation.

C: ANA5701，由于间隔，此转弯将引导你穿过航向道。

（13）　C: CXA1883, **you have crossed the localizer**, turn left immediately and return to the localizer.

C: CXA1883，你已穿过航向道，立即左转返回航向道。

（14）　C: UAL7429, **maintain 170 knots** until 7 miles from touchdown, contact Tower 118.1, good day.

C: UAL7429，保持 170 直到距接地点 7 海里，联系塔台 118.1，再见。

（15）　**P:** Dongfang Arrival, CES5121, heavy, maintaining 5100 meters, approaching VYK, Information C received.

C: CES5121, Dongfang Arrival, radar contact, Information C is valid, **vectoring for ILS approach**, Runway 36R, QNH 1007.

P: Roger, ILS approach, Runway 36R, QNH 1007, CES5121.

C: CES5121, turn left heading 350.

P: Left heading 350, CES5121.

C: CES5121, **report speed**.

P: Speed 250 knots, CES5121.

C: CES5121, **reduce to minimum clean speed**.

P: Reducing to 210 knots, CES5121.

C: CES5121, descend to 2100 meters on QNH 1006.

P: Descending to 2100 meters on QNH 1006, CES5121.

C: CES5121, turn right heading 090 for base.

P: Right heading 090, CES5121.

C: CES5121, **reduce to minimum approach speed**, turn left heading 030, **cleared for ILS approach**, **Runway** 36R, **report established**.

P: Reducing to minimum approach speed, left heading 030, cleared for ILS approach, Runway 36R, CES5121.

（A moment later）

P: CES5121, established.

C: CES5121, **no ATC speed restrictions**, position 15 kilometers from touchdown, **radar service terminated**, contact tower 118.8, good day.

P: 118.8, good day, CES5121.

P: 东方进场，CES5121，重型，5100 米保持，接近大王庄，通播 C 收到。

C: CES5121，东方进场，雷达看到了，通播 C 有效，雷达引导，ILS 进近，跑道 36 右，修正海压 1 007。

P: 收到，ILS 进近，跑道 36 右，修正海压 1 007，CES5121。

C: CES5121，左转航向 350。

P: 左转航向 350，CES5121。

C: CES5121，报告速度。

P: 速度 250，CES5121。

C: CES5121，减到最小光洁速度。

P: 减速到 210，CES5121。

C: CES5121，下降到修正海压 2 100 米，修正海压 1 006。

P: 下降到修正海压 2 100 米，修正海压 1 006，CES5121。

C: CES5121，右转航向 090 飞向四边。

P: 右转航向 090，CES5121。

C: CES5121，减到最小进近速度，左转航向 030，可以 ILS 进近，跑道 36 右，建立报告。

P: 减到最小进近速度，左转航向 030，可以 ILS 进近，跑道 36 右，CES5121。

（稍后）

P: CES5121，建立了。

C: CES5121，无 ATC 速度限制，距接地点 15 千米，雷达服务终止，联系塔台 118.8，再见。

P: 118.8，再见，CES5121。

（16）　**P:** Tianjin Approach, B3213.

C: B3213, Tianjin Approach.

P: B3213, Robinson R44, VFR from Tanggu helicopter airport to Binhai airport. 600 meters, estimating Binhai airport at 30, Information B.

P: 天津进近，B3213。

C: B3213，天津进近。

P: B3213，Robinson R44，目视飞行规则，从塘沽直升机机场飞往滨海国际机场。高度 600 米，预计到达滨海机场 30 分，通播 B 收到。

C: B3213, cleared to Binhai airport, VFR, QNH 1011.
P: Cleared to Binhai airport, VFR, QNH 1011, B3213.
C: B3213, report aerodrome in sight.
P: Wilco, B3213.
（A moment later）
P: B3213, aerodrome in sight.
C: B3213, contact Tower 118.1.
P: 118.1, B3213.

C: B3213，可以目视飞行到滨海国际机场，修正海压 1 011。
P: 可以目视飞行到滨海机场，修正海压 1 011，B3213。
C: B3213，能见机场报告。
P: 照办，B3213。
（稍后）
P: B3213，能见机场。
C: B3213，联系塔台 118.1。
P: 118.1，B3213。

（17） P: Dongfang Approach, CSN3356, request **straight-in** VOR approach, Runway 04.

P: 东方进近，CSN3356，请求直线 VOR 进近，跑道 04。

C: CSN3356, Dongfang Approach, cleared straight-in VOR approach, Runway 04.

C: CSN3356，东方进近，可以直线 VOR 进近，跑道 04。

（18） C: CCA1231, **are you familiar with** ECH01 Arrival?

C: CCA1231，你熟悉 ECH01 号进场程序吗？

练 习 题

1. 听力练习

说明：听两段 SIA802 航班的录音，完成表 5-1 和表 5-2 中所要求的内容。
（1）Dialogue 1

表 5-1

信息项	内容
STAR	
Information	
flight level to descend to	
QNH	
reduce speed to	
Final East frequency	

（2）Dialogue 2

表 5-2

信息项	内容
reduce speed to	
heading for base	
traffic clock	
traffic distance	

续表

信息项	内容
traffic landing runway	
heading for intercepting localizer	
landing runway	
Tower frequency	

2. 口语练习

说明：角色扮演，两人一组。根据对话内容，两人分别扮演管制员和驾驶员。管制员还需要根据通话内容和进近飞行进程单的格式（见图 5-1）及数据项说明（见表 5-3）[①]，完成进近飞行进程单（如图 5-2 所示）的填写。

TIME now is 1102.

P: CES2357, Dongfang Approach, 5400 meters maintaining, Information D.

C: CES2357, Dongfang Approach, radar contact, expect ILS approach, Runway 18, Information D is valid.

P: ILS approach, Runway 18, CES2357.

C: CES2357, descend to 3600 meters, report reaching.

P: Descending to 3600 meters, wilco, CES2357.

（A moment later）

P: CES2357, maintaining 3600 meters.

C: CES2357, descend to 900 meters on QNH 1011.

P: Descending to 900 meters on QNH 1011, CES2357.

（A moment later）

C: CES2357, turn right heading 210, cleared for ILS approach, Runway 18, report established.

P: Right heading 210, cleared for ILS approach, Runway 18, will report established, CES2357.

P: CES2357, established localizer.

C: CES2357, position 18 kilometers from touchdown, radar service terminated, contact Tower on 118.1, good day.

P: Contact Tower on 118.1, good day, CES2357.

（1）进近飞行进程单的格式如图 5-1 所示。

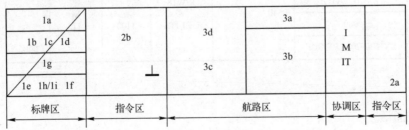

图 5-1

[①] 进程单格式及数据项说明参照《空中交通监视服务》中的附录 7 "飞行进程单使用方法"。

（2）数据项说明如表 5-3 所示。

表 5-3

数　据　项	代　表　内　容
标牌区	
1a	航空器呼号
1b	航空器机型
1c	尾流标志
1d	二次雷达应答机模式及编码
1e	起飞机场
1f	目的地机场
1g	巡航真空速（单位：公里/小时）
1h	预计（实际）起飞时刻
1i	预计降落时刻
/	表示已脱波航空器
指令区	
2a	申请的巡航高度层
2b	高度变化
⊥	已发布进近许可
航路区	
3a	位置报告点名称
3b	位置报告时刻
3c	进/离港程序
3d	起飞或落地跑道
协调区	
I	表示雷达识别，当航空器已识别，则在进程单"I"上画"○"
M	航空器已收到本场 ATIS 信息，则在进程单"M"上画"○"
IT	航空器已确认收到管制意图，则在进程单"IT"上画"○"

CES2357		GOL		
B747/H　A4516	0540	R/W18		I
900		GOL11A	1100	M IT
ZYHB0920/1130ZBDF				0980

图 5-2

3. 完成通话

说明：根据图 5-3 所示的进程单，已知 QNH 1 011，编写相应的通话内容并分角色练习，然后听录音核对答案。

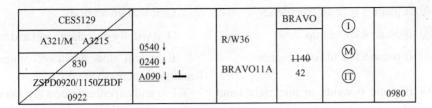

图 5-3

4. 听说练习

（1）说明：听录音，从下列航空器呼号中选择与图 5-4 中飞机图标所对应的呼号。

GCR6529   EVA780   CBJ5567   AMU082   CSZ9153   CSN3392

plane 1:_____,   plane 2:_____,

plane 3:_____,   plane 4:_____,

plane 5:_____,   plane 6:_____.

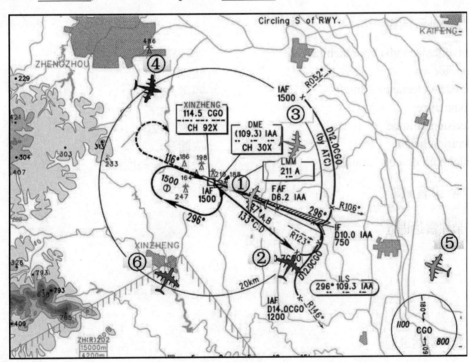

图 5-4

（2）说明：如图 5-4 所示，根据各航空器的位置，将下列左右两侧的内容组成意思表达正确的句子。

    ① If plane 1 can not land due to severe wind shear

    ② If plane 2 is too close to the preceding traffic

A. it will improve airspace efficiency.

B. it will be number 2 or 3 to land.

③ If plane 3 is in an emergency

④ If plane 4 speeds up

⑤ If plane 5 is a little bit higher

⑥ If plane 6 wants to join right hand downwind

C. it will go around.

D. it will fly over the airport and turn right.

E. it will slow down or continue present heading to increase the distance.

F. it will expedite descent to a lower level .

5.2　雷达进近

雷达进近主要有监视雷达进近和精密雷达进近两类，监视雷达进近通常作为一种应急程序。

监视雷达进近（surveillance radar approach，SRA）时，管制员为驾驶员提供航空器距离接地端的距离和建议高度，以便驾驶员能够在指定的跑道上实施进近。

精密雷达进近时，管制员不仅要为航空器提供航向信息，还需要提供高度信息。

5.2.1　单词与词组

surveillance radar approach (SRA)	监视雷达进近
touchdown	接地
obstacle clearance altitude	超障高度
commence	开始
decision altitude	决断高度
decision height	决断高
minimum descent altitude	最低下降高度

5.2.2　听录音 🔊

1. 监视雷达进近

2. 精密雷达进近

5.2.3　范例

（1）　**C:** CSN3106, Dongfang Precision, approaching glide path, **heading is good**.

C: CSN3106，东方精密，正在接近下滑道，航向好。

（2） **C: CCA1356, do not acknowledge further transmissions**, on track, approaching glide path.

C: CCA1356，请不要认收进一步发话，在航迹上，接近下滑道。

（3） **C: CAL512, check your minima.**

C: CAL512，检查最低标准。

（4） **C: CHB6263, check wheels down and locked.**

C: CHB6263，检查起落架放下锁好。

（5） **C: CUA5998, 5 1/2 miles from touchdown, altitude should be 2000 feet.**

C: CUA5998，距离接地点 5.5 海里，高度应为 2 000 英尺。

练 习 题

口语练习

说明：角色扮演，两人一组。两人分别扮演管制员和驾驶员，参照通话样例进行通话训练。

（1）监视雷达进近

例如：

P: Dongfang Radar, CCA1331.

C: CCA1331, Dongfang Radar, this will be a surveillance radar approach, Runway 27, terminating at 1/2 mile from touchdown, obstacle clearance altitude 400 feet, maintain 2200 feet, check your minima.

P: 2200 feet, Runway 27, CCA1331.

C: CCA1331, turn right heading 275 for final, report runway in sight.

P: Right heading 275, CCA1331.

C: CCA1331, approaching 6 miles from touchdown, commence descent now to maintain a 3 degrees glide path.

P: Descending, CCA1331.

C: CCA1331, check wheels down and locked.

P: Roger, CCA1331.

C: CCA1331, 5 1/2 miles from touchdown, altitude should be 2000 feet.

P: Roger, CCA1331.

C: CCA1331, going right of track, turn left heading 270.

P: Left heading 270, CCA1331.

C: CCA1331, 5 miles from touchdown, altitude should be 1900 feet.

P: Roger, CCA1331.

C: CCA1331, closing from the right, 4 1/2 miles from touchdown, altitude should be 1700 feet.

P: Roger, CCA1331.

C: CCA1331, wind calm, Runway 27, cleared to land.

P: Cleared to land, Runway 27, CCA1331.

C: CCA1331, 4 miles from touchdown, altitude should be 1600 feet, do not acknowledge further transmissions.

　　C: 3 1/2 miles from touchdown, altitude should be 1400 feet.

　　C: On track, turn right heading 272, 3 miles from touchdown, altitude should be 1300 feet.

　　C: 2 1/2 miles from touchdown, altitude should be 1100 feet.

　　C: 2 miles from touchdown, altitude should be 900 feet.

　　C: On track, heading is good, 1 1/2 miles from touchdown, altitude should be 800 feet.

　　C: On track, 1 mile from touchdown, approach completed.

（2）精密雷达进近

例如：

　　C: CSN3356, Dongfang Precision, report heading and altitude.

　　P: Heading 240 at 3000 feet, CSN3356.

　　C: CSN3356, position 6 miles east of Dongfang, turn right heading 260, descend to 2500 feet on QNH 1014.

　　P: Left heading 260, descending to 2500 feet on QNH 1014, CSN3356.

　　C: CSN3356, closing from the right, turn right heading 270.

　　P: Right heading 270, CSN3356.

　　C: CSN3356, approaching glide path, heading is good.

　　P: Roger, CSN3356.

　　C: CSN3356, how do you read?

　　P: Read you 5, CSN3356.

　　C: CSN3356, do not acknowledge further transmissions, on track approaching glide path.

　　C: Check your minima.

　　C: Commence descent now at 500 ft/min.

　　C: Check wheels down and locked.

　　C: On glide path, 5 miles from touchdown.

　　C: Turn right 5 degrees, new heading 275, I say again, 275.

　　C: 4 miles from touchdown, slightly below glide path.

　　C: Below glide path 100 feet, adjust rate of descent.

　　C: 50 feet below glide path, turn left heading 270, 3 miles from touchdown.

　　C: Coming back to the glide path.

　　C: On glide path, 2 1/2 miles from touchdown.

　　C: CSN3356, wind calm, Runway 27, cleared to land.

　　C: On glide path.

　　C: Heading 270 is good, slightly above glide path.

　　C: 2 miles from touchdown.

　　C: Coming back to the glide path.

　　C: On glide path, 1 3/4 miles from touchdown.

C: Turn right 2 degrees, new heading 272.

C: 1 1/2 miles from touchdown.

C: On glide path, 1 1/4 miles from touchdown, rate of descent is good.

C: On glide path, 1 mile from touchdown.

C: 3/4 of a mile from touchdown, on glide path.

C: 1/2 of a mile from touchdown, on glide path.

C: 1/4 of a mile from touchdown, approach completed.

本章综合练习

一、听力练习

说明：仔细观察图 5-5 中航空器的位置及标牌信息，然后听录音，将航空器呼号填写在图 5-5 中所对应的方框内横线处。

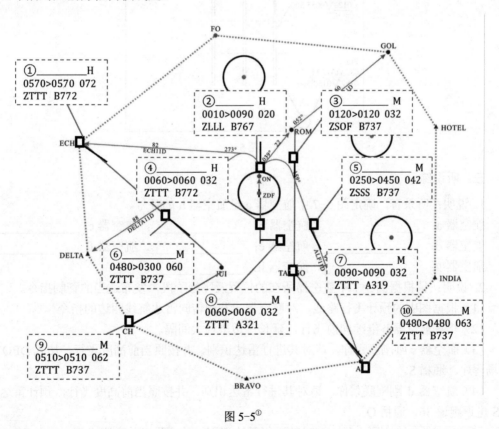

图 5-5①

二、完成通话

说明：如果你是进近管制员，如图 5-6 所示，按照下列要求给出通话内容。

---

① 空域图来自中国民航出版社《雷达管制实验指导》。

（1）CCA124 此时距离接地点 20 千米，指挥航空器继续进近，并联系塔台 118.6。

（2）根据航空器 CCA124 的当前速度和位置，向航空器 CSN378 发布速度调整指令。

（3）指挥 CCA1312 下降高度至 900 米并保持高度，发布 36L 跑道的进近许可指令。

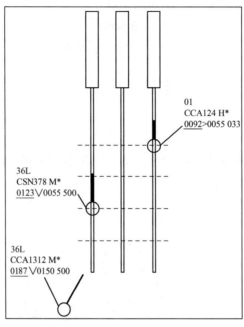

01
CCA124 H*
0092>0055 033

36L
CSN378 M*
0123∨0055 500

36L
CCA1312 M*
0187∨0150 500

图 5-6

三、听说练习

1. 说明：听录音，确定图 5-7 中航空器 a 至航空器 h 的呼号。

航空器 a： _____ ，航空器 b： _____ ，航空器 c： _____ ，

航空器 d： _____ ，航空器 e： _____ ，航空器 f： _____ ，

航空器 g： _____ ，航空器 h： _____ .

2. 说明：请根据图 5-7 中航空器所在的位置及飞行阶段，发布适当的管制指令。

（1）根据图中的预计飞行路线，对航空器 a 发布转向雷达航线四边的指令[①]。

（2）指挥航空器 b 继续向前飞行，延长三边来增大间隔。

（3）航空器 c 刚刚联系你，请对其进行雷达识别，并按照当前高度飞行，加入 DEO11 号进场程序，通播 S。

（4）航空器 d 刚刚联系你，请对其进行雷达识别，并按照当前高度飞行，预计雷达引导 ILS 进近跑道 36，通播 Q。

（5）航空器 e 刚刚联系你，请对其进行雷达识别，并按照当前高度飞行，加入 BRO11 进场程序，预计 ILS 进近跑道 36，通播 W。

---

① 雷达航线概念参见《空中交通监视服务》。

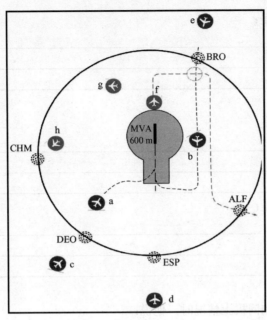

图 5-7

（6）航空器 f 刚刚离地，请对其进行雷达识别，指挥航空器保持一边方向飞行，上升到标准气压，4 500 米保持。

（7）指挥航空器 g 恢复自主领航，左转直飞 CHM。

（8）指挥航空器 h 联系东方区域 120.36。

四、思考题

说明：如图 5-7 所示，进场航空器 e 和离场航空器 f 存在高度穿越问题，管制员应发布何种指令才能避免两架航空器的冲突。

五、故事复述

说明：听录音，并复述故事内容。

六、小拓展：不守规矩的乘客（unruly passengers）

你听说过机上出现不守规矩的乘客的情况吗？听到航空器驾驶员有关这类乘客的报告时，管制员应该怎么处理呢？你将听到一段有关机上不守规矩的乘客的实况录音。听录音并回答以下问题。

（1）列出与不守规矩的乘客相关的单词或词组。

_____

_____

_____.

（2）当航空器驾驶员报告机上情况时，他报告了四次，第一次给出的信息是：

_____

_____

_____.

第二次给出的信息是：

_____

_____

_____.

第三次给出的信息是：

_____

_____

_____.

第四次给出的信息是：

_____

_____

_____.

（3）当管制员希望询问机上情况时，他说：

_____

_____

_____.

（4）当管制员希望了解骚乱的程度时，他说：

_____

_____

_____.

（5）当管制员希望知道不守规矩的乘客是否已被控制住时，他说：

_____

_____

_____.

第6章 机场管制——最后进近及着陆阶段

最后进近及着陆阶段包括航空器在五边建立稳定进近、着陆及滑入等常规运行阶段，还包括复飞程序及本场训练。最后进近和着陆阶段是航空器运行最为关键的阶段之一，此时驾驶员的操作、观察等工作负荷大增，因此，管制员标准、正确、适时的通话对保障航空器安全至关重要。此阶段的通话术语相对简练，涉及高度、速度及航向参数的指令不多，而是以"许可"为主。"许可"的结构相对简单，但发送时机非常重要。

通过本章的学习，应达到以下学习目标：

- ❖ 掌握起落航线飞行相关单词和用语；
- ❖ 掌握最后进近与着陆通话用语；
- ❖ 掌握复飞用语；
- ❖ 掌握脱离跑道、滑行入位等指令的格式和用语；
- ❖ 了解本场训练科目；
- ❖ 了解本场训练管制用语。

6.1 起落航线飞行

起落航线分为左起落航线和右起落航线。起飞后向左转弯建立的航线称为左起落航线，向右转弯建立的航线称为右起落航线，标准起落航线为左起落航线。起落航线通常由五条边、四个转弯（分别称为"一转弯""二转弯""三转弯""四转弯"）组成。如图 6-1 所示，假设航空器从跑道起飞，做一个左起落航线，航空器离地沿着跑道中心线延长线飞行，这个边叫"一边"（upwind），接着转向"二边"（crosswind），再做一个转弯转向"三边"（downwind），接下来转向"四边"（base），最后通过四转弯转向"五边"（final）准备着陆。

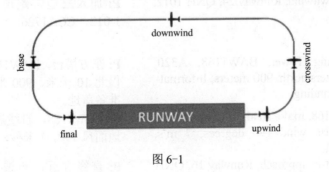

图 6-1

6.1.1 单词与词组

upwind   一边，顶风

turning crosswind	一转弯
crosswind	二边，侧风
turning downwind	二转弯
downwind	三边，顺风
turning base	三转弯
base	四边
turning final	四转弯
final	五边
turn to（crosswind, downwind, base, final）now	现在（一/二/三/四）转弯
join	加入
straight-in approach	直线进近
orbit	盘旋
make one orbit right	右盘旋一圈
make short approach	做小航线
extend downwind（make long approach）	延长三边（做大航线）

6.1.2 听录音 📢

6.1.3 范例

（1）　**P:** Dongfang Tower, CCA1726, A320, 10 kilometers north, 900 meters, request joining instructions to land.

C: CCA1726, **join downwind**, Runway 24, wind 270 degrees, 7 m/s, QNH 1012.
P: Join downwind, Runway 24, QNH 1012, CCA1726.

P: 东方塔台，CCA1726，A320，机场以北 10 千米，900 米保持，请求加入起落航线落地。

C: CCA1726，加入三边，跑道 24，地面风 270，7 米秒，修正海压 1 012。
P: 加入三边，跑道 24，修正海压 1 012，CCA1726。

（2）　**P:** Dongfang Tower, BAW1168, A320, 10 kilometers north, 900 meters, Information I, for landing.

C: BAW1168, **make straight-in approach**, Runway 16, wind 190 degrees, 7 m/s, QNH 1001.
P: Straight-in approach, Runway 16, QNH 1001, BAW1168.

P: 东方塔台，BAW1168，A320，机场以北 10 千米，900 米保持，通播 I，准备落地。

C: BAW1168，直线进近，跑道 16，地面风 190，7 米秒，修正海压 1 001。

P: 直线进近，跑道 16，修正海压 1 001，BAW1168。

（3）　P: Dongfang Tower, AMU866, downwind①.

C: AMU866, No. 2, follow B747 on base.

P: No.2, traffic in sight, AMU866.

P: Dongfang Tower, AMU866, base.

C: AMU866, **report final**.

P: Wilco, AMU866.
（A moment later）
P: Dongfang Tower, AMU866, final.

C: AMU866, **continue approach**, wind 270 degrees, 7 m/s.

P: 东方塔台，AMU866，三边。

C: AMU866，跟着四边的 B747，你是第二个落地。

P: 第二个，看到 B747 了，AMU866。

P: 东方塔台，AMU866，四边。

C: AMU866，五边报。

P: 照办，AMU866。
（稍后）
P: 东方塔台，AMU866，五边。

C: AMU866，继续进近，地面风 270，7 米秒。

（4）　C: CAL545, **extend downwind**, No. 2, follow A330 on 4 miles final.

P: No. 2, A330 in sight, CAL545.

C: CAL545, **make one orbit right** due traffic on the runway, report again on final.
P: Orbit right, CAL545.
（A moment later）
C: CAL545, **make short approach**, No.1.

P: Short approach, CAL545.

C: CAL545，延长三边，跟着五边 4 海里的 A330，你是第二个落地。

P: 第二个，看到 A330 了，CAL545。

C: CAL545，右转盘旋一圈，由于跑道上有活动五边报。

P: 右转盘旋圈，CAL545。
（稍后）
C: CAL545，做小航线，第一个落地。

P: 做小航线，CAL545。

练 习 题

1. 听力练习

说明：听录音，填空。

（1）**C**: CHB6263, _____.
　　　P: No. 2, B737 in sight, CHB6263.

（2）**C**: CSS467, _____.
　　　P: Orbit right, CSS467.

（3）**C**: CBJ5323, _____.
　　　P: Short approach, CBJ5323.

（4）**C**: B-7921, _____.
　　　P: Extend downwind, B-7921.

（5）**C**: _____.
　　　P: Join right hand downwind, Runway 36, QNH 1002, CSN3356.

---

① 已经加入起落航线的航空器，按照本场运行程序要求，驾驶员需要做常规的位置报告。

2. 听力练习

说明：听录音，在图6-2中起落航线适当的位置处标出各航空器。

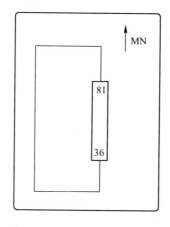

图6-2

3. 听说练习

说明：听录音，回答问题。

（1）Dialogue 1

① Which aircraft is calling Dongfang Tower?

② On which direction of the airfield is the aircraft?

③ Which position does the controller instruct the aircraft to join?

（2）Dialogue 2

① Which aircraft is calling Dongfang Tower?

② What type of approach is the aircraft going to make?

③ What is the wind like?

（3）Dialogue 3

① Which aircraft is calling Dongfang Tower?

② Where is the aircraft on the initial contact with the controller?

③ Which aircraft is No.1 to land? And where is it?

6.2　最后进近与着陆

　　最后进近中，当航空器转至五边时，距离接地地带在7千米（或4海里）之内，则报告"五边（final）"；距离接地地带大于7千米（或4海里），则报告"长五边（long final）"。如果航空器做直线进近，距离接地地带15千米（或8海里）时，航空器报告"长五边（long final）"。如果此时未收到着陆许可，则在距离接地地带7千米（或4海里）时，报告"五边（final）"。当航空器处于进近和着陆最后阶段时，除非出现紧急情况，不应向航空器发送指令或信息。

　　塔台管制员发出着陆许可后，如果条件变化，管制员必须立即通知航空器复飞，同时简要说明复飞原因。着陆或者复飞由驾驶员最后决定，并且对其决定负责。航空器复飞后，塔台管制员应为驾驶员提供复飞程序或航向高度，并将复飞航空器移交给进近管制员。

6.2.1　单词与词组

outer marker	外指点标
go around	复飞
missed approach	复飞（程序）
overshoot	飞越跑道
negative contact	没有看到
advise	告知，建议
minima	最低标准
wind shear	风切变

6.2.2　听录音 🔊

6.2.3　范例

（1）**P:** Dongfang Tower, UPS2921, long final.

　　C: UPS2921, Dongfang Tower, **continue approach**, report outer marker, wind 260 degrees, 7 m/s.
　　P: Continue approach, UPS2921.
　　（A moment later）
　　P: Dongfang Tower, UPS2921, final.
　　C: UPS2921, **wind 270 degrees, 7 m/s, Runway 27, cleared to land**.
　　P: Runway 27, cleared to land, UPS2921.

　　P: 东方塔台，UPS2921，长五边。

　　C: UPS2921，东方塔台，继续进近，过外指报告，地面风 260，7 米秒。

　　P: 继续进近，UPS2921。
　　（稍后）
　　P: 东方塔台，UPS2921，五边。
　　C: UPS2921，地面风 270，7 米秒，跑道 27，可以落地。
　　P: 跑道 27，可以落地，UPS2921。

（2）**C:** AAL128, Dongfang Tower, **go around**, aircraft on the runway.
　　P: Dongfang Tower, AAL128, going around.

　　C: AAL128，东方塔台，复飞，跑道上有飞机。
　　P: 东方塔台，AAL128，复飞。

（3）**P:** Dongfang Tower, ACA088, **no contact at minima, going around**.
　　C: ACA088, roger, report downwind.

　　P: 东方塔台，ACA088，最低高度无法能见地面，复飞。
　　C: ACA088，收到，三边报。

（4）**C:** CPA347, the broken airplane will be towed out of the runway in time, **continue approach, prepare for possible go-around**.

　　C: CPA347，损坏的飞机将被及时拖出跑道，继续进近，做好可能复飞的准备。

练 习 题

1. 口语练习

说明：角色扮演，两人一组。首先由最后进近的驾驶员联系管制员，报告航空器位置，管制员发布继续进近指令和着陆许可。然后交换角色再次练习。例如：

> **P:** Dongfang Tower, <u>CCA1331</u>, long final.
>
> **C:** <u>CCA1331</u>, Dongfang Tower, continue approach.
>
> **P:** Continue approach, <u>CCA1331</u>.
>
> **C:** <u>CCA1331</u>, wind <u>200 degrees</u>, <u>7 m/s</u>, Runway <u>18R</u>, cleared to land.
>
> **P:** Runway <u>18R</u>, cleared to land, <u>CCA1331</u>.

从表 6-1 中选取航空器呼号、跑道号及风向、风速进行替换后练习。

表 6-1

航空器呼号	跑 道 号	风向、风速
UAL889	18R	190 degrees, 6 m/s
SAS996	18L	220 degrees, 20 knots
DLH721	36L	340 degrees, 5 m/s, gusting to 10 m/s
CCA931	36R	wind calm

2. 听力练习

说明：听录音，填空。

（1）**P:** Dongfang Tower, CSH9129, outer marker.

　　C: _____.

　　P: Wilco, CSH9129.

　　（A moment later）

　　P: CSH9129, short final.

　　C: _____.

　　P: Runway 36L, cleared to land, CSH9129.

（2）**P:** Dongfang Tower, CSN3356, final.

　　C: CSN3356, Dongfang Tower, you are No.1 to land. _____

_____.

　　P: No.1 to land, roger, CSN3356.

　　（A moment later）

　　P: CSN3356, going around.

　　C: CSN3356, _____, when passing 1000 feet, turn right to JUI VOR.

（3）**C:** CES5323, _____

_____.

　　P: CES5323, we'll see at our _____.

　　（A moment later）

P: CES5323, going around.

C: CES5323, _____.

P: CES5323, what is the standard procedure?

C: _____.

P: Roger, 900 meters, contact 123.1, CES5323.

3. 口语练习

说明：角色扮演，两人一组。首先由驾驶员报告航空器位置，塔台管制员向航空器发布复飞指令，再分别根据图 6-3 和图 6-4 所示向驾驶员发布详细的复飞程序指令。然后交换角色再次练习。例如：

P: Beijing Tower, CSZ9886, outer marker.

C: CSZ9886, go around immediately, unreported vehicle on the runway.

P: Going around, CSZ9886.

C: CSZ9886, standard procedure, contact Beijing Approach on 121.65.

P: CSZ9886, what is the standard procedure?

C: (图 6-3) _____.

　　or

C: (图 6-4) _____.

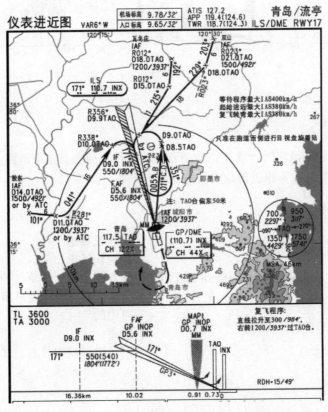

图 6-3

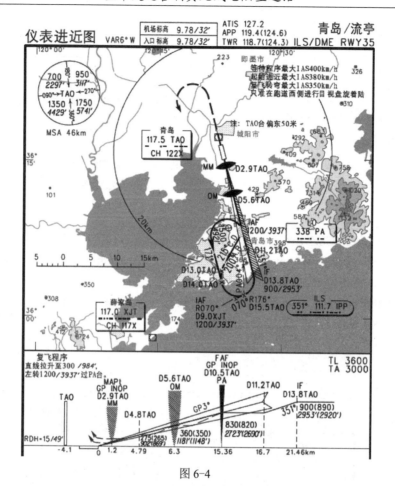

图 6-4

6.3　滑入

　　当航空器处于着陆滑跑时，除非完全必要，管制员不应在其结束滑跑前发布滑行指令。一般情况下，航空器在脱离跑道前，应保持在塔台频率上。

6.3.1　单词与词组

fast turnoff	快速脱离道
exit	跑道快速出口
take first (or second…) convenient left (or right)	前方第一个（或第二个……）合适的道口左（或右）转脱离跑道
available	可用的
vacate	脱离
take next left(right)	下一个道口左（右）转脱离

6.3.2　听录音 🔊

6.3.3 范例

（1）　**C1:** HDA841, **vacate runway via Taxiway E.**

　　P: Vacate runway via Taxiway E, HDA841.

　　（A moment later）

　　P: Dongfang Tower, HDA841, runway vacated.

　　C1: HDA841, contact Dongfang Ground on 121.7, good day.

　　P: 121.7, HDA841, good day.

　　P: Dongfang Ground, HDA841, runway vacated.

　　C2: HDA841, Dongfang Ground, **report follow-me car in sight.**

　　P: Follow-me car in sight, HDA841.

　　C2: HDA841, follow the follow-me car, goodbye.

　　P: Follow the follow-me car, goodbye. HDA841.

C1: HDA841，沿 E 滑行道脱离跑道。

P: E 滑行道脱离跑道，HDA841。

（稍后）

P: 东方塔台，HDA841，已脱离跑道。

C1: HDA841，联系东方地面 121.7，再见。

P: 121.7，HDA841，再见。

P: 东方地面，HDA841，已脱离跑道。

C2: HDA841，东方地面，见引导车报告。

P: 看到引导车了，HDA841。

C2: HDA841，跟着引导车滑行，再见。

P: 跟着引导车滑行，HDA841，再见。

（2）　**C1:** AAL8901, **expedite vacating** runway, aircraft on short final. **When vacated, contact** Dongfang Ground on 119.2, goodbye.

　　P: Expedite vacating, Dongfang Ground 119.2, AAL8901, goodbye.

　　（A moment later）

　　P: Dongfang Ground, AAL8901, runway vacated.

　　C2: AAL8901, Dongfang Ground, **hold position, give way to** B747 entering Taxiway H.

　　P: Holding, AAL8901.

　　（A moment later）

　　C2: AAL8901, taxi to Stand B213 via Taxiway H, D2 and D3, good day.

　　P: Stand B213 via Taxiway H, D2 and D3, AAL8901, good day.

C1: AAL8901，加速脱离跑道，短五边有落地飞机。脱离后，联系东方地面 119.2，再见。

P: 加速脱离，东方地面 119.2，AAL8901，再见。

（稍后）

P: 东方地面，AAL8901，已脱离跑道。

C2: AAL8901，东方地面，原地等待，给正在滑入 H 滑行道的 B747 让路。

P: 等待，AAL8901。

（稍后）

C2: AAL8901，沿滑行道 H、D2 和 D3 滑到停机位 B213，再见。

P: 沿 H、D2 和 D3 滑到停机位 B213，AAL8901，再见。

（3）　**C:** ANA8489, **take first right, report runway vacated**[①].

C: ANA8489，前方第一个道口右转，脱离后报告。

---

① 通常在低能见度下使用。

P: First right, report runway vacated, ANA-8489.

　　（A moment later）

P: Dongfang Tower, ANA8489, runway vacated.

C: ANA8489, taxi to Gate 12 via Taxiway C, C3 and C4. **Caution, construction work in progress** adjacent to Gate12.

P: Gate 12 via Taxiway C, C3 and C4, ANA8489.

P: 第一个道口右转，脱离后报告，ANA8489。

　　（稍后）

P: 东方塔台，ANA8489，脱离跑道。

C: ANA8489，沿滑行道 C、C3 和 C4 滑到停机位 12。注意，停机位 12 旁正在施工。

P: 沿滑行道 C、C3 和 C4 滑到停机位 12，ANA8489。

练 习 题

1. 口语练习

说明：角色扮演，两人一组。两人分别扮演管制员和驾驶员进行着陆后滑行通话，然后交换角色再次练习。例如：

C: HDA331, take the first convenient left turnoff, when vacated, contact Ground 121.7.

P: 121.7, HDA331.

　　（A moment later）

P: HDA331, runway vacated.

C: HDA331, taxi to Stand 18 via A4 and C3.

P: Via A4 and C3, Stand 18, HDA331.

从表 6-2 中选取航空器呼号、脱离道、地面频率、滑行道及停机位进行替换后练习。

表 6-2

航空器呼号	脱 离 道	地面频率	滑 行 道	停 机 位
QFA129	take A1 fast turnoff	122.1	G, L	Gate 14
CSN3356	take first left	119.7	A4, C3	Stand 18
ANA963	backtrack, exit 4	122.6	B2, D3	Gate D4
DAL185	take first left turnoff	122.5	2nd intersection K	Stand B5

2. 口语练习

说明：请根据机场停机位置（如图 6-5 所示）及表 6-3 中的信息（着陆跑道为 17 或 35），为航空器发送滑行指令。例如：

C: CES546, taxi to Gate 7 via Taxiway A, A2.

P: Gate 7 via Taxiway A, A2, CES546.

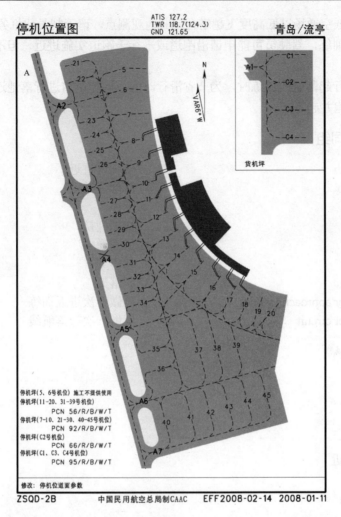

图 6-5

　　各航空器的落地信息如表 6-3 所示，根据各航空器对应的着陆跑道和停机位信息，确定其所使用的滑行路线，并为各航空器发送滑行指令。

表 6-3

航空器呼号	着陆跑道	停 机 位
CAL501	35	23
EVA712	17	10
SAS856	35	33
GCR7843	17	38

6.4　本场训练

　　常见的本场训练的形式包括低空通场、低高度进近和落地连续等形式。

　　当驾驶员需要地面帮助目视检查航空器状态时，如起落架是否收放好，起落架舱门是否关闭等，可申请低空通场以低高度飞越塔台或其他观测点，请管制员或机务人员检查。

为了做进近训练，驾驶员可能申请沿跑道或平行于跑道实施进近，但不着陆，这种训练叫作低高度进近。

当驾驶员进行起落航线训练时，为节省滑行时间，可能申请进行落地连续，即着陆之后继续进行起飞滑跑并起飞。

6.4.1 单词与词组

low approach	低高度进近
touch and go	落地连续
full stop	全停
congestion	交通拥挤
low pass	低空通场
landing gear	起落架
undercarriage	起落架
make another approach training	再做一次进近训练
make another circuit	再做一次起落航线

6.4.2 听录音 🔊

1. 低空通场

2. 低高度进近

3. 落地连续

6.4.3 范例

（1）P: Dongfang Tower, AHK8783, our nose gear is jammed, request low pass to have the undercarriage checked.
C: AHK8783, Dongfang Tower, **cleared low pass**, Runway 18, **not below 150 meters**, report final.
P: Runway 18, not below 150 meters, report final, AHK8783.

P: 东方塔台，AHK8783，由于前起落架卡住了，请求低空通场进行目视检查。
C: AHK8783，东方塔台，可以低空通场，跑道 18，不低于 150 米，五边报。
P: 跑道 18，不低于 150 米，五边报，AHK8783。

（2）C: JOY1517, **landing gear appears down**.

C: JOY1517，起落架看起来放下了。

（3）　P: Dongfang Tower, YZR7991, request low **approach**, Runway 23 for training.

C: YZR7991, Dongfang Tower, **cleared low** approach, Runway 23, not below 150 meters, report final.

P: Runway 23, not below 150 meters, report final, YZR7991.

P: 东方塔台，YZR7991，请求低高度进近训练，跑道23。

C: YZR7991，东方塔台，可以低高度进近，跑道23，不低于150米，五边报。

P: 跑道23，不低于150米，五边报，YZR7991。

（4）　P: Dongfang Tower, N231CL, **request touch and go**.

C: N231CL, Dongfang Tower, cleared touch and go.

P: Cleared touch and go, N231CL.

P: 东方塔台，N231CL，请求落地连续。

C: N231CL，东方塔台，可以落地连续。

P: 可以落地连续，N231CL。

（5）　P: Dongfang Tower, CSN6624, request touch and go.

C: CSN6624, Dongfang Tower, negative due traffic congestion, **make full stop**, Runaway 36, cleared to land.

P: Runaway 36, cleared to land, full stop, CSN6624.

P: 东方塔台，CSN6624，请求落地连续。

C: CSN6624，东方塔台，不同意，由于交通拥挤，做全停，跑道36，可以落地。

P: 跑道36，可以落地，做全停，CSN6624。

练 习 题

1. 口语练习

说明：角色扮演，两人一组。首先由驾驶员联系管制员请求做低高度进近，管制员发布许可。然后交换角色再次练习。例如：

> P: Dongfang Tower, CCA1331, request low approach for training, Runway 16.
>
> C: CCA1331, Dongfang Tower, cleared low approach, Runway 16, not below 150 meters on QNH 998, report final.
>
> P: Runway 16, not below 150 meters on QNH 998, CCA1331.

从表 6-4 中选取航空器呼号和跑道号进行替换后练习。

表 6-4

航空器呼号	跑 道 号
UAL889	36 R
SAS996	18 L
DLH723	18 R
CSC8974	36 L

2. 口语练习

说明：角色扮演，两人一组。首先由驾驶员联系管制员请求做低空通场，管制员发布许可。然后交换角色再次练习。例如：

> **P:** DongfangTower, <u>SBI101</u>, unsafe left gear indication, request low pass near the Tower to have the undercarriage checked.
>
> **C:** <u>SBI101</u>, DongfangTower, cleared low pass Runway <u>36 L</u>, not below 150 meters on QNH 1 004, north of Tower.
>
> **P:** Runway <u>36 L</u>, not below 150 meters on QNH 1004, north of Tower, <u>SBI101</u>.

从表 6-5 中选取航空器呼号和跑道号进行替换后练习。

表 6-5

航空器呼号	跑 道 号
DKH1254	23
CAL501	32
KZR888	01
CKK206	19

3. 口语练习

说明：角色扮演，两人一组。首先由驾驶员联系管制员请求做落地连续，管制员发布许可。然后交换角色再次练习。例如：

> **P:** Tower, AFR123, Stand 24, request local flight training. We need two circuits, one touch and go, one full stop.
>
> **C:** AFR123, roger, stand by.
>
> （Tower calls Approach.）
>
> **C:** AFR123, local flight training approved, one touch and go, one full stop.
>
> **P:** Roger, AFR123.
>
> **C:** AFR123, line up and wait, Runway 35.
>
> **P:** Line up and wait, Runway 35, AFR123.
>
> **C:** AFR123, when airborne, join left hand traffic circuit Runway 35. Surface wind 310 degrees, 7 m/s, cleared for take-off.
>
> **P:** When airborne, join left hand traffic circuit, Runway 35, cleared for take-off, wind 310 degrees, 7 m/s, AFR123.
>
> （A moment later）
>
> **P:** Final, AFR123.
>
> **C:** AFR123, surface wind 310 degrees, 7 m/s, Runway 35, cleared touch and go. Report downwind.
>
> （AFR123 is about to make a full stop this time.）

C: AFR123, surface wind 310 degrees, 7 m/s, Runway 35, cleared to land for full stop.

6.4.4　听力练习

说明：听录音，填空。

（1）**P:** Dongfang Tower, CKK206, request touch and go.

　　C: CKK206, Dongfang Tower, _____

　　_____.

　　P: Cleared to land for full stop, CKK206.

（2）**P:** Dongfang Tower, G-CHDG, request touch and go.

　　C: G-CHDG, Dongfang Tower, _____.

　　P: Runway 36, cleared touch and go, G-CHDG.

（3）**P:** BAW259, outer marker, request touch and go for training.

　　C: BAW259, _____

　　_____.

　　P: Roger, cleared touch and go, will call you downwind, BAW259.

　　C: BAW259, _____

　　_____.

　　P: Going around, BAW259.

（4）**P:** Dongfang Tower, CSN5013, request low pass, _____.

　　C: CSN5013, Dongfang Tower, _____.

　　P: Runway 27, not below 150 meters, CSN5013.

　　（A moment later）

　　C: CSN5013, _____.

　　P: Roger, will advise intentions.

（5）**P:** Dongfang Tower, GCR7843, request low approach, Runway 09 for training.

　　C: GCR7843, Dongfang Tower, _____.

　　P: Runway 09, not below 150 meters, report final, GCR7843.

本章综合练习

一、词汇练习

说明：请将下列单词按照重音位置归类。

holding　　terminal　　negative　　expediting　　intention

destination　　traffic　　identified　　direction

●●	●●●	●●●	●●●●	●●●●

二、词汇练习

说明：请将下列单词按照-ed 的读音归类。

reduced　　　landed　　　taxied　　　vacated　　　turned

missed　　　confused　　　awaited　　　realized　　　instructed

/id/	/t/	/d/

三、词汇练习

说明：请将左右两列的单词搭配，形成具有特定含义的短语。

1. braking　　　　　　　　a. gate
2. leading　　　　　　　　b. personnel
3. parking　　　　　　　　c. flights
4. warning　　　　　　　　d. action
5. boarding　　　　　　　　e. point
6. outbound　　　　　　　f. bay
7. ground handling　　　　g. edge
8. holding　　　　　　　　h. light

四、口语练习

说明：两人一组，根据图 6-6 所示的信息进行对话。学生 A 向学生 B 发布航空器着陆和滑行指令，学生 B 听到学生 A 的指令后在图上标出滑行路线或航空器位置，学生 A 核对。然后互换角色，再次练习。

五、完成通话

说明：根据英文提示，完成管制员通话，并听录音核对答案。

（1）**P:** Dongfang Tower, CSN3520, we've established ILS approach Runway 22.

（Give CSN3520 landing clearance, the surface wind is on 210 degrees at 4 m/s.）

C: _____.

P: Runway 22, Cleared to land, CSN3520.

（2）（You find that CSN3520 is above the glide path on the radar screen. Verify his altitude.）

C: _____.

P: Dongfang tower, the PAPI indicates we are a little bit above, going around.

（3）（Tell CSN3520 to climb to 1200 meters, fly heading 250, and then call Departure on 124.35.）

C: _____.

（4）（You instruct HVN105 to vacate runway at the first convenient right and change to Ground 121.6.）

C: _____.

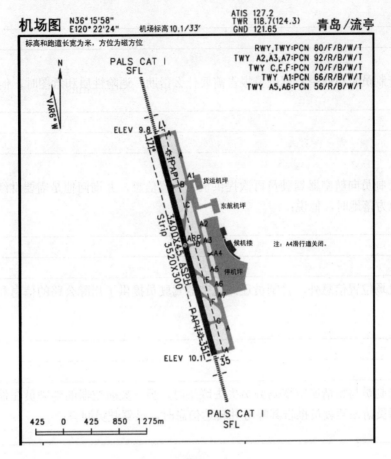

图 6-6

（5）（You instruct JAL635 to taxi via Taxiway C, C1 to Stand D3.）

　　　　C: _____.

六、故事复述

说明：听录音，并复述故事内容。

七、小拓展：如何在无线电陆空通话中更好地沟通？

作为空中交通管制（学）员，你是否认识到无线电陆空通话中沟通不畅的不良影响？怎样才能实现良好沟通呢？你将听到一段实况录音，看看管制员如何处理发生的特情。听录音并回答以下问题。

（1）当管制员希望证实机上情况时，他说：

_____

_____

_____.

他还说：

_____

_____

_____ .

（2）当管制员希望航空器驾驶员报告需要什么帮助、遇险性质和意图时，他说：

_____

_____

_____ .

（3）当管制员向航空器驾驶员再次提供机场位置信息，并询问他是需要继续前往机场还是另寻其他地方落地时，他说：

_____

_____

_____ .

（4）除机场位置信息外，管制员还向航空器驾驶员提供了州际公路的信息和多次障碍物信息：

_____

_____

_____ .

（5）当管制员与特情航空器驾驶员失去联络时，另一架航空器的驾驶员在他们之间转达信息，当管制员请该驾驶员报告其呼号并转达信息时，该驾驶员回答：

_____

_____

_____ .

（6）管制员希望跟转达信息的航空器驾驶员保持联络，并告诉驾驶员如果由于某些原因失去联络时，该驾驶员可以自行转换到下一管制单位频率，他说：

_____

_____

_____ .

第7章 紧急情况

　　紧急情况下的通话与常规通话不同，往往没有明确规定的术语，更多以"明语"进行表达。通常以国际上比较习惯、通用的表达作为参考。通话时，管制员应使用镇定、清楚、明确、自信的语音，放慢语速，避免不必要的重复。

通过本章的学习，应达到以下学习目标：

❖ 掌握遇险及紧急情况的发布与认收用语；
❖ 掌握遇险及紧急情况的处置与取消用语；
❖ 了解特殊情况的类别。

　　常见紧急情况如表 7-1 所示。

表 7-1

英　文	中　文
bomb threat	炸弹威胁
control system failure	操纵系统故障
power system failure	动力系统故障
total electrical failure	电力系统故障
engine failure	发动机故障
de-icing/anti-icing system failure	除冰/防冰系统故障
windshield（英式为 windscreen）problems	风挡问题
fire in the hold/toilet/cabin	货舱起火/厕所起火/客舱起火
hijack	劫机
radar failure	雷达失效
under carriage failure	起落架故障
fuel problems	燃油问题
communication failure	通信失效
hydraulic system failure	液压系统故障
passenger with a heart attack/ injuries among passengers and cabin crew after severe turbulence/pregnant woman	患心脏病的乘客/严重颠簸导致乘客和空乘人员受伤/孕妇
instrument failure	仪表故障
pressurization system failure/decompression (depressurization)	增压系统故障/失压
bird strike	鸟击
low visibility	低能见度
wind shear	风切变
thunderstorm/icing/turbulence	雷暴/积冰/颠簸

7.1　发布与认收

根据航空器出现问题的严重程度，紧急情况可以分为遇险和紧急。

遇险（distress）是指航空器或航空器上人员遇到迫在眉睫的危险的严重威胁，需要立即得到救助的状态。如果航空器处于遇险状态，驾驶员在给管制员报告时，通常会以"MAYDAY"来开始通话，遇险信号应重复三次，即"MAYDAY, MAYDAY, MAYDAY"。遇险信号比所有通话都具有优先权。

紧急（urgency）是指航空器或航空器上人员的安全受到威胁，但不需要立即援助的一种状态。如果航空器处于紧急状态，驾驶员在给管制员报告时，通常会以"PANPAN"来开始通话，紧急信号应重复三次，即"PANPAN, PANPAN, PANPAN"。除遇险信号以外，紧急信号比其他所有通话都具有优先权。

航空器运行过程中，如果出现了遇险情况，驾驶员一般会按照表7-2所示的格式向管制员发布信息。

表 7-2

英 文 格 式	中 文 格 式
① MAYDAY, MAYDAY, MAYDAY	① MAYDAY, MAYDAY, MAYDAY
② ATC unit	② 接收电台的名称
③ aircraft identification	③ 航空器识别标志
④ nature of the distress	④ 遇险的性质
⑤ intention of the person in command and request	⑤ 驾驶员意图和请求
⑥ position, level and heading of the aircraft	⑥ 航空器位置、高度和航向
⑦ any other useful information	⑦ 其他有用信息

航空器运行过程中，如果出现了紧急情况，驾驶员一般会按照表7-3所示的格式向管制员发布信息。

表 7-3

英 文 格 式	中 文 格 式
① PANPAN, PANPAN, PANPAN	① PANPAN, PANPAN, PANPAN
② ATC unit	② 接收电台的名称
③ aircraft identification	③ 航空器识别标志
④ nature of the urgency	④ 紧急的性质
⑤ intention of the person in command and request	⑤ 驾驶员意图和请求
⑥ position, level and heading of the aircraft	⑥ 航空器位置、高度和航向
⑦ any other useful information	⑦ 其他有用信息

7.1.1　单词与词组

MAYDAY　　　　　　　　　　　　　遇险的话呼
PANPAN　　　　　　　　　　　　　紧急的话呼
port　　　　　　　　　　　　　　左（舷/翼）

starboard	右（舷/翼）
ditch	水上迫降
extinguish	扑灭
dump fuel	放油
burn off	消耗
roger MAYDAY	收到 MAYDAY
intoxicated	醉酒
radioactive	放射性的

7.1.2 听录音 🔊

7.1.3 范例

（1）　**C:** CKK206, Dongfang Approach, **roger MAYDAY/PANPAN**.

　　　C: CKK206，东方进近，收到 MAYDAY/PANPAN。

（2）　**P:** CSH9070, position POU, emergency descent to 3600 meters due to decompression.
　　　C: Attention, all aircraft in the vicinity of POU, **emergency descent in progress** from 8900 meters to 3600 meters.

　　　P: CSH9070，位置平洲，由于失压，紧急下降到 3 600 米。
　　　C: 平洲附近的所有航空器请注意，有航空器正从 8 900 米紧急下降到 3 600 米。

（3）　**P:** G-BDES, **transmitting blind due to receiver failure**. G-BDES 7500 meters, heading 110, over CHG VOR this time, descending to be at 3600 meters over GOL, standard arrival procedure, next for landing Runway 36 at Dongfang.

　　　P: G-BDES，接收机失效，盲发。G-BDES，高度 7 500 米，航向 110，CHG 上方，下降到 3 600 米通过 GOL，标准进场程序，东方机场跑道 36 落地。

（4）　**C:** All aircraft, Dongfang Control, **fuel dumping in progress**, on radial 230 JUI VOR, ranging from 15 to 20 miles, avoid flight below 600 meters within 10 miles of the fuel dumping track.

　　　C: 全体注意，东方区域，有航空器正在放油中，JUI 230 度径向线，范围 15 到 20 海里，避免进入放油航迹下方 600 米，水平 10 海里范围内。

练　习　题

1. 听力练习

说明：听录音，根据驾驶员的紧急呼叫，写出航空器所遭遇的问题和（或）驾驶员的请求。

（1）_____.

（2）_____.

（3）_____.

（4）_____.

（5）_____.

2. 口语练习

说明：结对活动。根据表 7-4 和表 7-5 所给信息，练习询问遇险性质和机长意图。学生 A 和学生 B 分别扮演管制员和驾驶员，然后交换角色再次练习。例如：

C: State the nature of your emergency.
P: We are experiencing problems with our No.1 engine.
C: What are your intentions?
P: We'd like to return to the airport for landing.

Student A

表 7-4

Problem	Course of Action
engine fire and No.1 engine shutdown	we're going to return to the airport for landing
passenger is having a heart attack	we will need to divert to an airport with medical staff
the visibility is too low	we're going to divert to our alternate

Student B

表 7-5

Problem	Course of Action
intoxicated passenger who has assaulted a flight attendant	we're going to return to the airport for landing
we've got a large thunderstorm over the airport	we will enter a holding pattern
problems with our landing gear	we are going to make a low approach, so you can visually inspect it

7.2 处置与取消

遇险（或紧急）呼叫通常应在当前使用的频率上完成。如果波道内通话繁忙，管制员认为将遇险航空器移交到另一频率将有利于问题的解决，可以实施通话强制静默；如果将其他航空器移交到另一频率，而专心在当前频率协助遇险航空器更有效，则可指挥其他航空器联系另一频率。当管制员得知遇险结束，应通知所有相关电台。

7.2.1 单词和词组

stop transmitting   停止通信

cancel distress	取消遇险
distress traffic ended	遇险活动结束
rear	后部，尾部
smoke	烟
cargo hold	货舱
fire truck	消防车
dangerous goods	危险品
persons on board	机上人数

7.2.2　听录音 🔊

7.2.3　范例

（1）　C: **All stations**, Dongfang Approach, **stop transmitting, MAYDAY**.

　　　C: 全体注意，东方进近，停止发话，MAYDAY。

（2）　C: MAYDAY, G-BCDO, Dongfang Approach, contact Control on 127.1.

　　　C: MAYDAY，G-BCDO，东方进近，联系区域 127.1。

（3）　P: Dongfang Approach, G-BCDO, **cancel distress**, engine restarted.
　　　C: MAYDAY, all stations, Dongfang Approach, time 1325, **distress traffic G-BCDO ended**.

　　　P: 东方进近，G-BCDO，取消遇险，发动机重启了。
　　　C: MAYDAY 全体注意，东方进近，1325 分，G-BCDO 遇险活动结束。

练 习 题

1. 听力练习

说明：听录音，回答问题。

（1）What problem did AIC349 experience?

（2）What did the pilot request for?

（3）What might happen after landing according to the pilot?

（4）What information did the controller ask for?

2. 听力练习

说明：听录音，填空。

（1）P: CPA748, we've _____.

　　C: Do you have any other _____? Did you see the _____ or the _____?

　　P: It was a _____, that's all we know.

　　C: _____?

P: Affirm. It was a very close thing. ＿＿＿＿＿＿＿＿＿＿＿＿＿＿＿＿＿＿＿＿.

（A moment later）

P: CPA748, ＿＿＿＿＿＿ passengers have been ＿＿＿＿＿＿＿＿＿＿, but there's a doctor ＿＿＿＿＿＿＿＿＿＿, so we'll continue on our ＿＿＿＿＿＿.

C: Roger, CPA748.

（2）**P:** ALK866, we have a ＿＿＿＿＿＿＿＿＿＿＿＿＿＿＿, request ＿＿＿＿＿＿ to Dongfang.

C: ALK866, ＿＿＿＿＿＿＿＿＿＿＿＿＿＿＿＿＿＿＿＿＿＿＿＿＿＿＿.

P: Turning right heading 280, descending to 3600 meters, ALK866.

C: ＿＿＿＿＿＿＿＿＿＿＿＿＿＿＿＿＿＿＿＿＿＿＿＿＿＿＿＿＿＿.

P: Affirm, ALK866.

C: Roger, will ＿＿＿＿＿＿.

（3）**C:** JAL7908, Dongfang Control, your company ＿＿＿＿＿＿＿＿＿＿ us that ＿＿＿＿＿＿ ＿＿＿＿＿＿＿＿＿＿＿＿＿＿＿＿＿＿＿＿＿＿＿＿.

P: Do you have any ＿＿＿＿＿＿ about ＿＿＿＿＿＿＿＿＿＿＿＿＿＿＿＿＿＿＿?

C: Negative.

P: ＿＿＿＿＿＿＿＿＿＿ to Nanping Airport, request ＿＿＿＿＿＿＿＿＿＿＿＿＿＿＿＿＿＿, JAL7908.

本章综合练习

一、听力练习

说明：听录音，请选出听到的航空器航班号，在对应的字母前画"✓"。

（1）a. 171　　　　　　b. 979　　　　　　c. 179

（2）a. 425　　　　　　b. 429　　　　　　c. 494

（3）a. 686　　　　　　b. 868　　　　　　c. 668

（4）a. 6939　　　　　b. 6393　　　　　c. 6339

（5）a. 4213　　　　　b. 2413　　　　　c. 4321

（6）a. 5797　　　　　b. 5791　　　　　c. 9579

（7）a. 4081　　　　　b. 4031　　　　　c. 4051

（8）a. 7951　　　　　b. 5791　　　　　c. 7195

（9）a. 6338　　　　　b. 6388　　　　　c. 6833

（10）a. 1255　　　　b. 1295　　　　c. 5129

二、听力练习

说明：听五段录音，分别回答下列问题。

（1）Why did the pilot shut down the engine?

（2）Why did the aircraft come back?

（3）What should the crew do before returning to Nanping?

（4）Why did the aircraft divert?

（5）What problem did the crew encounter?

三、听力练习

说明：听一段录音，回答下列问题。

（1）What problem did QTR5844 experience?

（2）What did the pilot decide to do?

（3）Draw a picture to represent the flight path of QTR5844.

四、口语练习

说明：询问具体信息。学生 A 读出 Student A 上的句子，并就画线部分提问，学生 B 回答。然后学生 B 读出 Student B 上的句子，并就画线部分提问，学生 A 回答。例如：

Student A：

The two aircraft were controlled by the <u>Lambourne Control Centre</u>.

学生 A 问学生 B：Who was controlling the two aircraft?

学生 B 回答：The Lambourne Control Center.

Student B：

The <u>two aircraft</u> were controlled by the Lambourne Control Centre.

学生 B 问学生 A：How many aircraft were controlled?

学生 A 回答：Two aircraft.

Student A

（1）The captain is throttling back <u>to slow down the aircraft</u>.

（2）Center told us <u>to stay at this altitude</u> until advised.

（3）I'm advising <u>air traffic</u> that we'll need to hold.

（4）We've got more than <u>3 hours</u> of fuel remaining.

（5）The LAM holding pattern <u>is based on a VOR</u> near Romford.

（6）There were a lot of delays <u>due to single-runway operation</u> at Heathrow.

（7）At 1638 an MD81 entered the pattern and was instructed to <u>maintain 6300 meters</u>.

（8）The B737 had <u>approximately 5 miles</u> to go to the LAM VOR.

（9）The MD-81 was heading <u>west</u> inbound to LAM.

（10）The MD81 was 300 meters <u>directly below</u> the B737.

Student B

（1）The captain <u>is throttling back</u> to slow down the aircraft.

（2）Center told us to stay at this altitude <u>until advised</u>.

（3）I'm advising air traffic <u>that we'll need to hold</u>.

（4）We've got more than 3 hours of <u>fuel remaining</u>.

（5）The LAM holding pattern is based on a VOR <u>near Romford</u>.

（6）There were <u>a lot of delays</u> due to single-runway operation at Heathrow.

（7）<u>At 1638</u> an MD81 entered the pattern and was instructed to maintain 6300 meters.

（8）The B737 had approximately 5 miles to go <u>to the LAM VOR</u>.

（9）The MD81 was heading west inbound to LAM.

（10）The MD81 was 300 meters directly below the B737.

五、故事复述

说明：听录音，并复述故事内容。

六、小拓展：ASSIST 原则

在非正常或紧急情况下，管制员可以遵循 ASSIST 原则，为航空器驾驶员提供各种便利、信息和情报，协助其尽快转危为安。请大家一起谈谈对 ASSIST 原则的认识。

ASSIST 原则由 Acknowledge、Separate、Silence、Inform、Support、Time 六个单词的首字母组成，其中：

（1）Acknowledge—Make sue you understood the nature of emergency and acknowledge accordingly.

_____

_____

_____.

（2）Separate—Don't forget to establish/maintain separation!

_____

_____

_____.

（3）Silence—Impose silence on your control frequency if necessary. Don't disturb urgent cockpit actions by unnecessary transmissions!

_____

_____

_____.

（4）Inform—Inform your supervisor and other sectors/units concerned.

_____

_____

_____.

（5）Support—Give maximum support to pilot and crew.

_____

_____

_____.

（6）Time—Allow pilots sufficient time to work on their problem.

_____

_____

_____.

附录 A "听录音"原文

第2章 机场管制——起飞前与起飞阶段

2.1 无线电检查

P: Dongfang Ground, CSN3401, radio check on 121.5. How do you read?

C: CSN3401, Dongfang Ground, readability 5.

2.2 离场条件

2.2.1 离场信息

P: Pudong Ground, CXA101, request departure information.

C: CXA101, Pudong Ground, departure Runway 17, wind 180 degrees, 4 m/s, QNH 1 015, temperature 28, dew point 21, RVR 550 meters.

2.2.2 机场通播

Beijing Capital Airport Information D, 0930 UTC, ILS approach, Runway 18R, wind 300 degrees, 5 m/s, CAVOK, temperature 28, dew point 25, QNH 1010, no significant other information. On initial contact advise you have Information D.

2.3 重要机场情报

C: UPS102, Dongfang Tower, runway surface condition, Runway 36L, centerline lighting unserviceable.

2.4 放行许可

P: Dongfang Delivery, CSH9070, destination Shanghai, with Information C, request ATC clearance.

C: CSH9070, Dongfang Delivery, cleared to Shanghai via flight planned route, ALPHA11D Departure, initial climb to 900 meters on QNH 1013, request level change for 9500 meters en route, squawk 2013, after departure, contact Approach 119.1.

P: Cleared to Shanghai via flight planned route, ALPHA11D Departure, initial climb to 900 meters on QNH 1013, request level change for 9500 meters en route, squawk 2013, after departure, contact Approach 119.1, CSH9070.

C: CSH9070, read-back correct.

2.5 推出开车

P: Beijing Ground, CSN303, Stand A5, request push-back and start up.

C: CSN303, push-back and start up approved.

2.6 滑出

P: Beijing Ground, CAL501, request taxi.

C: CAL501, Beijing Ground, taxi to holding point K2, Runway 36R via Taxiway K, hold short of Runway 36R.

2.7 起飞

C: CSN2105, Dongfang Tower, line up, Runway 18L.

P: Lining up, Runway 18L, CSN2105.

C: CSN2105, wind 240 degrees, 10 m/s, Runway 18L, cleared for take-off.

P: Cleared for take-off, Runway 18L, CSN2105.

第 3 章　进近管制——离场阶段

3.1 雷达管制用语

C: CPA295, confirm squawking 3426.

P: Affirm, CPA295.

C: CPA295, radar contact, maintain 9800 meters, continue present heading.

P: Roger, maintaining 9800 meters, continue present heading, CPA295.

C: CPA295, turn left heading 210.

P: Left heading 210, CPA295.

3.2 离场指令

P: Dongfang Departure, CCA1352, airborne at 18, passing 130 meters climbing to 900 meters.

C: CCA1352, Dongfang Departure, radar contact, follow YV 51 Departure, continue climb to 2700 meters on QNH 1011.

P: Climbing to 2,700 meters on QNH 1,011,YV 51D Departure, CCA1352.

C: CCA1352, continue climb to 5400 meters on standard, climb at 2000 ft/min or greater.

P: Climbing to 5400 meters at 2000 ft/min or greater, CCA1352.

    (A moment later)

P: CCA1352, maintaining 5400 meters.

C: CCA1352, roger.

C: CCA1352, contact Dongfang Control on 120.1, good day.

P: 120.1 for Control, CCA1352, good day.

3.3 飞行活动通报

C: CSN3369, Dongfang Control, unknown traffic, 2 o'clock, 10 kilometers, crossing right to left.

P: Looking out, CSN3369. Request vectors.

C: CSN3369, avoiding action. Turn right 30 degrees immediately and report heading.

P: Turning right 30 degrees, heading 090, CSN3369.

C: CSN3369, now clear of traffic. Resume own navigation to XFA.

第4章 区域管制

4.1 高度信息

P: Beijing Control, CCA1331, over HUR, maintaining 4500 meters.

C: CCA1331, Beijing Control, radar contact, climb to and maintain 6300 meters.

P: Climbing to 6300 meters, CCA1331.

C: CCA1331, continue climbing to 8900 meters, report reaching.

P: Climbing to 8900 meters, CCA1331.

(A moment later)

P: CCA1331, approaching 8900 meters.

C: CCA1331, maintain 8900 meters, contact Dongfang Control on 123.4, good day.

P: Dongfang Control, 123.4, good day.

4.2 位置信息

P: Dongfang Control, CCA4378, ECH at 35, maintaining 10,400 meters, estimating MEDAL at 48.

C: CCA4378, roger, next report at COAST.

4.3 航路等待

C: CCA1331, Dongfang Control, cleared to MKE, maintain 7500 meters, hold inbound track 180 degrees, left hand pattern, outbound time one and a half minutes, expect further clearance at 17.

P: Roger, hold at MKE, inbound track 180 degrees, outbound time one and a half minutes, left hand pattern, CCA1331.

4.4 RVSM 运行与 SLOP

C: AAR311, Dongfang Control, confirm RVSM approved.

P: Dongfang Control, AAR311, Negative RVSM.

C: AAR311, unable issue clearance into RVSM airspace, maintain 8400 meters.

P: Maintaining 8400 meters, AAR311.

4.5 绕飞雷雨

P: Beijing Control, CCA1730, we have an indication of severe weather 30 kilometers ahead. Request to turn right to go round it.

C: CCA1730, Beijing Control. Negative. Due to the restricted area, turn left 30 degrees and track out 35 kilometers, report clear of the build-ups.

P: Turning 30 degrees left and 35 kilometers out, CCA1730.

（A moment later）

P: Beijing Control, CCA1730, we're clear of Cbs now.

C: Roger, CCA1730, turn right heading to come back on track.

4.6　航空器加入、穿越或离开航路

1. 加入航路

P: Dongfang Control, B1213.

C: B1213, Dongfang Control.

P: B1213, request clearance to join G17 at ANDIN.

C: B1213, cleared to Nanyuan, flight planned route, 3600 meters. Join G17 at ANDIN at 3600 meters.

P: Cleared to Nanyuan via ANDIN, flight planned route, 3600 meters. To enter controlled airspace 3600 meters, B1213.

C: B1213, read-back correct.

2. 穿越航路

P: Dongfang Control, B1716.

C: B1716, Dongfang Control.

P: B1716, Yun-7, 20 miles north of KIG VOR, 2400 meters, KIG VOR at 33, request clearance to cross airway V21 at KIG VOR.

C: B1716, cleared to cross V21 at KIG VOR, 2400 meters.

P: Cleared to cross V21 at KIG VOR, 2400 meters, B1716.

C: B1716, report at KIG VOR.

P: Wilco, B1716.

3. 离开航路

C: B2341, Dongfang Control, cleared to leave A1 via JUI VOR. Maintain 4500 meters while in controlled airspace.

P: Cleared to leave A1 via JUI VOR. Maintaining 4500 meters while in controlled airspace, B2341.

第 5 章　进近管制——进场阶段

5.1　进场及进近

P: Beijing Arrival, CXA8101, 5100 meters maintaining, Information W received.

C1: CXA8101, Beijing Arrival, radar contact. Follow JB7A RNAV arrival, expect radar vector for ILS approach, Runway 36L, Information W is valid.

P: ILS approach, Runway 36L, CXA8101.

C1: CXA8101, descend to 3600 meters on standard.

P: Descending to 3600 meters on standard, CXA8101.

C1: CXA8101, descend to 1800 meters on QNH 1006.

P: Descending to 1800 meters on QNH 1006, CXA8101.

C1: CXA8101, contact Beijing Final 126.1, good day.

P: Final 126.1, good day, CXA8101.

    (A moment later)

P: Beijing Final, CXA8101, 1800 meters maintaining.

C2: CXA8101, Beijing Final, landing Runway 36L.

P: Runway 36L, CXA8101.

C2: CXA8101, reduce speed to 180 knots.

P: Reducing speed to 180 knots, CXA8101.

C2: CXA8101, turn right heading 020, cleared for ILS approach, Runway 36L, report established on localizer.

    P: Right heading 020, ILS approach, Runway 36L, will report established, CXA8101.

    (A moment later)

P: CXA8101, established localizer, Runway 36L.

C2: CXA8101, continue approach, Runway 36L, contact Tower 124.3, good day.

P: Continue approach, Runway 36L, 124.3, good day, CXA8101.

5.2 雷达进近

1. 监视雷达进近

C: CCA1501, Dongfang Radar, this will be a surveillance radar approach, Runway 18, terminating at 1/2 mile from touchdown, obstacle clearance altitude 500 feet, maintain 2,000 feet, check your minima.

2. 精密雷达进近

C: CSN3106, Dongfang Precision, approaching glide path, heading is good.

P: Dongfang Precision, CSN3106.

C: CSN3106, do not ackonwlege further transmissions, on track, approaching glide path.

C: CSN3106, check your minima.

C: CSN3106, commence descent now at 500 feet.

C: CSN3106, check wheels down and locked.

第6章 机场管制——最后进近及着陆阶段

6.1 起落航线飞行

P: Dongfang Tower, B-7321, Yun-7, 10 kilometers north of the field, 900 meters, request joining instructions to land.

    C: B-7321, Dongfang Tower, join downwind, Runway 36, wind 150 degrees, 10 m/s, QNH 1,015.

P: Join downwind, Runway 36, QNH 1015, B-7321.

6.2　最后进近与着陆

C: CES5412, Dongfang Tower, wind calm, Runway 36L, cleared to land.
P: Runway 36L, cleared to land, CES5412.

6.3　滑入

C1: CCA1320, Dongfang Tower, wind 110 degrees, 5 m/s, Runway 09, cleared to land.
P: Cleared to land, Runway 09, CCA1320.

　　(A moment later)

C1: CCA1320, take first left, when vacated, contact Ground 118.35.
P: First left, 118.35, CCA1320.

　　(A moment later)

P: Dongfang Ground, CCA1320, runway vacated.
C2: CCA1320, Dongfang Ground, taxi to Stand 28 via Taxiway A and C.
P: Taxi to Stand 28 via Taxiway A and C, CCA1320.

6.4　本场训练

1. 低空通场

P: Dongfang Tower, B2711, request low pass for landing gear check.
C: B2711, Dongfang Tower, cleared low pass, Runway 18R.

2. 低高度进近

P: Dongfang Tower, B3644, request low approach and overshoot for another approach training.
C: B3644, Dongfang tower, cleared low approach, Runway 18L.

3. 连续起飞

P: Dongfang Tower, G-ABCD, request touch and go.
C: G-ABCD, Dongfang Tower, cleared touch and go.
P: Cleared touch and go, G-ABCD.

第7章　紧急情况

7.1　发布与认收

P: MAYDAY MAYDAY MAYDAY, Dongfang Tower, G-BYCM, DC-10, port engine on fire. Position 15 kilometers south of Dongfang, 1200 meters, losing altitude. Heading 010. Request straight-in approach to Dongfang.
C: G-BYCM, Dongfang Tower, roger MAYDAY.

7.2　处置与取消

P: MAYDAY, MAYDAY, MAYDAY, Dongfang Control, D-BIXT. We have fire in the rear toilet. We're descending to 4500 meters. Request an emergency landing at Dongfang. Position 50 miles west of Dongfang. Heading 075.

C: D-BIXT, Dongfang Control, roger MAYDAY. Break, break, all stations on 126.3, stop transmitting, MAYDAY.

(A moment later)

P: MAYDAY. D-BIXT. Fire now under control. Cancel distress.

C: Roger, D-BIXT. MAYDAY, all stations, distress traffic ended.

附录 B　练习题录音原文及答案

第1章　基本运行程序

1.1　概述

2. 完成通话

答案:

（1）**P:** Dongfang Control, CES2105.

　　C: CES2105, Dongfang Control.

　　C: CES2105, descend to 9500 meters.

　　P: Descending to 9500 meters, CES2105.

（2）**P:** Dongfang Approach, BAW038.

　　C: BAW038, Dongfang Approach.

　　C: BAW038, climb to 6300 meters.

　　P: Climbing to 6300 meters, BAW038.

1.2　字母

3. 听力练习

（1）LATIC	124.25
（2）TNODP	134.2
（3）DIKTU	121.32
（4）WITYO	128.9
（5）JAITU	FL 210
（6）GLKIR	7200 meters
（7）SHIRC	Heading 235
（8）IEHOB	Runway 35
（9）DRAWI	wind 240 degrees, 12 knots
（10）KSCBY	1340

1.3　数字

3. 听力练习

录音原文及答案：

（1）BAW038, climb to <u>7200 meters</u> on standard.

（2）HDA903, descend to <u>3000 meters</u> on QNH 1011.

（3）CCA1108, climb to <u>5100 meters</u> on standard.

（4）CES2105, descend to <u>2700 meters</u> on QFE 1014.

（5）CDG4670, descend to <u>3000 meters</u> on QNH 1010.

6. 口语练习

答案：

（b）turn left heading 330.

（c）turn right heading 020.

（d）turn right heading 330.

（e）turn left heading 020.

（f）turn right heading 090.

7. 听力练习

录音原文及答案：

（1）CCA1108, turn right heading <u>180</u> for base, Runway 27, report visual.

（2）CES2105, turn left heading <u>080</u> for downwind leg.

（3）KLM4302, turn right heading <u>300</u> to intercept the localizer. Report established.

（4）CPA347, turn left heading <u>360</u>, descend to 4 500 meters.

（5）CCA1356, turn right heading <u>240</u>.

10. 听力练习

录音原文及答案：

（1）CSN3102, Dongfang Control, contact Dongfang Approach <u>118.1</u>, good day.

（2）CHH7805, contact Wuhan Tower <u>130.0</u>.

（3）Dongfang Approach, CDG4670, request change to <u>127.1</u>.

（4）HDA890, monitor ATIS <u>123.250</u>.

（5）CPA903, contact Dongfang Control <u>121.35</u>.

13. 听力练习

（1）squawk <u>1342</u>.

（2）squawk <u>5364</u>.

（3）squawk <u>2743</u>.

（4）squawk <u>6352</u>.

（5）squawk <u>4345</u>.

14. 听力练习

录音原文及答案：

（1）BAW038, report the <u>Airbus 321</u> on final.

（2）UAL889, give way to the <u>Boeing 747</u> on your left side.

（3）CES2105, traffic, 11 o'clock, 6 miles, <u>MA60</u>, crossing left to right.

（4）CCA1108, go around, an <u>Airbus 320</u> is stuck on the runway.

（5）CSN3102, follow the <u>EMB145</u> to Taxiway A.

15. 听力练习

录音原文及答案：

（1）CSN301, Dongfang Tower, <u>visibility 5000 meters</u>.

（2）CES3108, Dongfang Tower, <u>RVR Runway 36L 3000 meters</u>.

（3）KLM898, Dongfang Tower, <u>RVR Runway 18R not reported.</u>

（4）UAL889, Dongfang Tower, <u>RVR Runway 36R touchdown 2000 meters, midpoint 2200 meters, stop end 2100 meters</u>.

（5）CCA101, Dongfang Tower, <u>cloud broken, cumulus, 900 meters</u>.

1.4 标准单词和词组

2. 完成通话

（1）acknowledge

（2）approved

（3）cleared

（4）break

（5）break break

（6）contact

（7）confirm

（8）correction, negative, read-back correct

（9）I say again

（10）Affirm/Negative

4. 听力练习

录音原文：

（1）**C:** CES9070, climb to 2500 feet.

   P: Climbing 22500 feet, CES9070.

   C: CES9070, confirm climbing to 2500 feet.

   P: Affirm, climbing to 2500 feet, CES9070.

（2）P: Hong Kong Radar, good afternoon, JAL020, FL 140 on the 104 radial.

　　C: JAL020, good afternoon, remain on the 104 radial, climb to FL 330.

　　P: Climbing to FL 330, maintain the 140 radial, JAL020.

　　C: JAL020, confirm maintaining the 104 radial.

　　P: Affirm we will remain the 104 radial, JAL020.

（3）P: Hong Kong Radar, HDA903, what speed would you like on the descent?

　　C: HDA903, maintain 300 knots or less.

　　P: 310 knots, HDA903.

　　C: HDA903, that is 300 knots, confirm.

　　P: Roger, 300 knots, HDA903.

（4）C: CPA347, reduce speed to 310 knots.

　　P: Reduce speed to 230 knots, CPA347.

　　C: CPA347, confirm reducing to 310 knots.

　　P: Affirm reducing to 310 knots, CPA347.

（5）C: QFA129, descend to FL 150, leave CH, heading 180.

　　P: Descending to 1500, leave CH, heading 180, QFA129.

　　C: QFA129, just confirm descend to FL 150.

　　P: Affirm 150, QFA129.

答案:

表 1-27

驾驶员以为听到的是	管制员实际说的是
（1）climb to 22500 feet	climb to 2500 feet
（2）remain on the 140 radial	remain on the 104 radial
（3）310 knots	maintain 300 knots or less
（4）reduce speed to 230 knots	reduce speed to 310 knots
（5）descend to 1500	descend to FL 150

5. 思考题

答案:

（1）语言问题，方言影响，文化背景，相似呼号等。

（2）例如: 英文通话中的"to"和"two"，"wet"和"wait"，"fire"和"via"，"proceeding"和"preceding"，"left"和"right"。中文通话中的"1"和"7"。

1.5　呼号

1.5.1　管制单位呼号

答案:

（1）Dongfang Delivery, BAW038, request ATC clearance.

（2）CCA1108, Dongfang Approach, cleared for ILS approach.

（3）CES2105, Dongfang Control, descend to 9500 meters.

（4）Dongfang Tower, DKH1254, request taxi.

（5）CXA8130, Dongfang Departure, climb to 4800 meters.

1.5.2　航空器呼号

2. 完成通话

录音原文及答案：

（1）**P:** Beijing Approach, CES5144, Information I, 5400 meters maintaining.

    C: CES5144, maintain 5400 meters. Report passing HUR.

    P: Maintain 5400 meters. Will report passing HUR, CES5144.

（2）**P:** Guangzhou Approach, CSN3168.

    C: CSN3168, Guangzhou Approach, report heading and level.

    P: Heading 060, 4500 meters, CSN3168.

（3）**C:** CCA1375, descend to and maintain 2400 meters on QNH 1004.

    P1: Descend to and maintain 2400 meters on QNH 1004, CCA1357.

    C: CCA1357, negative, maintain present level. That was for CCA1375.

    P1: Maintain present level, CCA1357.

    C: CCA1375, Dongfang Approach, descend to and maintain 2400 meters on QNH 1004.

    P2: Descend to and maintain 2400 meters on QNH 1004, CCA1375.

    C: CCA1375, read-back correct.

3. 听力练习

答案：

CHH7761	QTR803	JAL883	CDG4676
HVN6416	CES5263	UAL858	HDA811
SIA831	CDG4646	KAL893	CSN3554
JAL891	CQH8876	CES2668	CSH9206
QTR807	UAL835	CSC8954	CES5623
CSN3524	CSH9260	JAL81	HVN416
UAE309	CSC8974	CQH8867	CHH7721

1.6　通信

1.6.1　通信的建立

答案：

P: Dongfang Delivery …1331.

C: Station calling Dongfang Delivery, say again your call sign.

P: Dongfang Delivery, CCA1331.

C: CCA1331, Dongfang Delivery, stand by.

P: Standing by, CCA1331.

   (A moment later)

C: CCA1331, Dongfang Delivery.

P: Dongfang Delivery, CCA1331, <u>destination</u> Shanghai, Gate 5, request ATC clearance.

C: CCA1331, <u>confirm</u> you have got information C.

P: <u>Negative</u>, CCA1331.

C: CCA1331, <u>monitor</u> ATIS 128.2, call me when ready.

P: Monitoring 128.2, CCA1331.

   (A moment later)

P: Dongfang Delivery, CCA1331, <u>Information</u> C, request ATC clearance.

C: CCA1331, Cleared to Shanghai via <u>flight planned route</u>, ALPHA02 Departure, <u>initial climb</u> to 900 meters on QNH 1011, maintain 9500 meters on standard, <u>squawk</u> 3475.

P: Cleared to Shanghai <u>via</u> flight planned route, ALPHA02 Departure, initial climb to 900 meters on QNH 1011, maintain 9500 meters on standard, squawk 3475, CCA1331.

C: CCA1331, <u>read-back correct</u>, <u>contact</u> Dongfang Tower on 118.1.

P: 118.1, CCA1331.

1.6.2 许可的发布与复诵要求

答案:

（1）**C:** DLH 723, <u>turn left heading 120</u>.

    P: Turn left heading 130, DLH 723.

    C: Negative, turn left heading 120.

    P: <u>Turn left heading 120</u>, DLH 723.

    C: DLH 723, read-back correct.

（2）**C:** HDA903, <u>contact Dongfang Control 127.0</u>.

    P: Contact Dongfang Control 127.1.

    C: HDA903, negative, contact Dongfang Control on 127.0.

    P: <u>127.0, HDA903.</u>

    C: HDA903, read-back correct.

1.6.3 通信的移交

1. 听力练习

录音原文及答案:

（1）**C:** KLM570, contact <u>Pudong Tower</u> on <u>118.4</u>. Correction <u>118.8</u>.

    P: <u>118.8</u>, KLM570.

（2）**C:** UAL781, contact <u>Beijing Control</u> on <u>125.25</u>. Correction <u>125.35</u>.

    P: <u>125.35</u>, UAL781.

2. 听力练习

录音原文及答案:

（1）THA435, contact <u>Shanghai Control on 135.7</u>.

（2）CSN3243, contact <u>Beijing Control 118.1</u>.

（3）CXA1201, contact <u>Shanghai Approach 123.4</u>.

（4）DLH902, contact <u>Guangzhou Approach 127.4</u>.

（5）THY021, contact <u>Tianjin Approach 123.2</u>.

（6）CES2343, contact <u>Wuhan Control 122.3</u>.

（7）UAE309, landing time 23, take first right. Contact <u>Ground on 121.9</u>.

（8）AFR802, position 18 kilometers from touchdown. Contact <u>Tower on 118.4</u>.

（9）AAR3365, contact <u>Beijing Approach 119.2</u>.

（10）SWR801, continue approach, contact <u>Dongfang Tower 119.7</u>.

（11）AAL186, after departure, continue straight ahead and contact <u>Dongfang Approach on 124.35</u>.

本章综合练习

四、听力练习
答案：

（1）11300 meters

（2）121.45

（3）Runway 23

（4）FL 270

（5）QNH 1013

（6）Heading 260

（7）wind 320 degrees, 10 knots

（8）5000 feet

（9）119.75

（10）9800 meters

五、听力练习
答案：

（1）climb and maintain 10400 meters.

（2）contact Beijing Control 118.1.

（3）next report BHJ 14.

（4）change your call sign to BCHG.

（5）temperature 24, dew point 22.

（6）CCA1572, heavy, request taxi.

（7）Beijing Approach, CCA1243, information H.

（8）taxi to holding point Runway 23.

（9）radio check 123.45.

（10）turn right heading 340.

六、故事复述

录音原文：

On March 3, 1974, Turkish Airlines Flight 981, a McDonnell Douglas DC-10, crashed in a forest northeast of Paris, France. The London-bound plane crashed shortly after taking off from Orly airport. All 346 people on board died. It was later determined that the cargo door detached, which caused an explosive decompression; this caused the floor just above to collapse. The collapsed floor severed the control cables, which left the pilots without control of the elevators, the rudder and No. 2 engine. The plane entered a steep dive and crashed. It was the deadliest plane crash of all time until the Tenerife disaster in 1977.

七、小拓展：相似航班号。

参考答案：

（1）将呼号改为注册号；

（2）注意语音语调和适当停顿，重点突出相似航班号的差异部分；

（3）先呼出航班，再进行通话；

（4）中英文分别叫；

（5）在其中一架航班后加字母如 A。

第 2 章　机场管制——起飞前与起飞阶段

2.1　无线电检查

1. 听力练习

录音原文及答案：

（1）**P:** Dongfang Ground, CCA1356, radio check <u>118.1</u>.

　　C: CCA1356, Dongfang Ground, <u>read you 3</u>, you signal is <u>weak</u>, check your <u>transmitter</u>.

　　　（A moment later）

　　P: Dongfang Ground, CCA1356, 1, 2, 3, 4, 5, <u>how do you read now</u>?

　　C: CCA1356, <u>loud and clear</u>.

（2）**P:** Dongfang Ground, SIA802, radio check <u>129.0</u>. How do you read me?

　　C: Station calling Ground, <u>you are cutting in and out</u>, check your transmitter and give me a short count.

（3）**P:** Dalian Ground, CQH631, radio check <u>box 1</u> on <u>121.75</u>.

　　C: CQH631, Dalian Ground, readability <u>2</u>, <u>loud background whistle</u>.

3. 完成通话

答案：

P: Dongfang Ground, KLM889, radio check 121.6.

C: Station calling Dongfang Ground, say again your call sign.

P: Dongfang Ground, KLM889, radio check 121.6, how do you read me?

C: KLM889, read you 2.

P: Dongfang Ground, KLM889, call you back in 5 minutes.

P: Dongfang Ground, KLM889, 1,2,3,4,5, 5,4,3,2,1, how do you read me now?

C: KLM889, readability 5.

2.2 离场条件

2.2.1 离场信息

1. 听力练习

录音原文及答案：

（1）**P:** Dongfang Tower, AFR496, <u>destination</u> Paris, request <u>departure</u> information.

C: AFR496, Dongfang Tower, <u>runway in use 18R</u>, wind <u>250 degrees</u>, <u>12</u> m/s, <u>gusting to</u> 20 m/s, <u>visibility</u> more than 10 kilometers, temperature 25, dew point <u>19</u>, QNH <u>998</u>.

P: Runway 18R, QNH <u>998</u>, AFR496.

（2）**P:** Beijing Ground, SIA3310, <u>request departure information</u>.

C: SIA3310, Beijing Ground, <u>departure Runway 36L</u>, wind <u>290 degrees</u>, <u>4</u> m/s, QNH 1011, temperature <u>minus 2</u>, dew point <u>minus 6</u>, RVR <u>550 meters</u>.

P: Runway 36L, QNH 1 011, SIA3310.

（3）**P:** Hong Kong Ground, HDA101, request departure information.

C: HDA101, departure Runway <u>13</u> ,wind <u>160 degrees</u>, <u>10</u> knots, <u>gusting</u> 25 knots, QNH <u>1023</u>, temperature <u>27</u>, dew point 21, taxiway <u>B4</u> closed due <u>work in progress</u>.

P: Roger, Runway 13, QNH 1023, HDA101.

2.2.2 机场通播

1. 词汇练习

答案：

（1）light, moderate, heavy 描述雨的大小；

（2）damp, wet, standing water, flooded 描述跑道潮湿程度；

（3）less than poor, poor medium to poor, medium, good to medium, good 描述刹车效应；

（4）sky clear, few, scattered, broken, overcast 描述云量.

2. 听力练习

录音原文及答案：

This is Heathrow departure information <u>P</u>. <u>1620</u> hours, weather: <u>200 degrees</u>, <u>12</u> knots, temperature <u>plus 21</u>, dew point <u>plus 16</u>. QNH <u>1016</u>, departure Runway <u>27R</u>.

3. 听力练习

录音原文：

This is Orly at time 1310. Information I, ILS approach, landing Runway 07, take-off Runway 08, expect 3L Departure, braking action good, transition level 50. Caution, flock of birds east of airfield. Wind 080 degrees, 18 knots, visibility 700 meters, ceiling 4 octas 3000 feet, 6 oktas 7800 feet. Temperature 0, dew point minus 2, QNH 1009, QFE threshold 07 996, threshold 08 998. Advise first contact you have received information I.

答案:

ATIS　　I　, Time 1310 , Runway 07 for landing; 08 for take-off,

SID　3L Departure, TL 50.

4. 听力练习

录音原文:

All stations, the new information is B. The changes: wind 310 degrees, 10 to 15 knots, cloud few at 3000 feet, temperature 11, QNH 1004.

答案:

表 2-4

信　息　项	选　　　项			
ATIS designator	O	A	R	B
wind direction	310 degrees	360 degrees	160 degrees	60 degrees
wind speed	10–15 knots	15–35 knots	15–25 knots	25–35 knots
cloud amount	scattered	few	broken	overcast
cloud height	1000 feet	2000 feet	3000 feet	4000 feet
temperature	−1	−11	21	11
QNH	1004	1114	1104	1040

2.3　重要机场情报

1. 词汇练习

答案:

●●	●●●	●●●	●●●●	●●●●
hangar towing failure	terminated incident	retracted reported	information intersection	unserviceable available

2. 听力练习

录音原文及答案:

（1）**C**: AAL241, taxi with caution, work in progress on Taxiway 2.

　　　P: Roger, AAL241.

（2）**C**: MAS3351, taxi to holding point Runway 08. Be advised midpoint of taxiway partly covered by ice. Braking action medium.

　　　P: Roger, taxi to holding point Runaway 08, MAS3351.

（3）**C:** CHH801, taxi to holding point <u>Runway 14</u>. <u>Green centreline lights on taxiway</u> <u>unserviceable.</u>

    P: Roger. <u>Taxi to holding point Runway 14</u>, CHH801.

3.完成通话

答案：

（1）**C:** CSN1233, <u>caution construction work adjacent to Gate 36.</u>

    P: Roger, CSN1233.

（2）**C:** KLM898, <u>caution work in progress ahead north side of Taxiway A marked by red flag.</u>

    P: Roger, KLM898.

（3）**C:** SAS995, <u>caution construction vehicle just off Taxiway A.</u>

    P: Roger,SAS995.

（4）**C:** CDG332, <u>caution Taxiway B centerline lighting unserviceable.</u>

    P: Roger, CDG332.

（5）**C:** LKE278, wind calm, Runway 09, cleared for take-off. <u>Caution flock of birds sighted</u> <u>1 kilometer east of airfield.</u>

    P: Runaway 09, cleared for take-off, LKE278

2.4 放行许可

1.听力练习

录音原文及答案：

（1）**P:** Pudong Delivery, CCA1502, Gate 12 to Beijing, request ATC clearance.

    C: CCA1502, Pudong Delivery, cleared to Beijing, PIKAS2D Departure, <u>initial climb and</u> <u>maintain 900 meters</u>, <u>departure frequency 120.3, squawk 4501</u>.

（2）**P:** Pudong Delivery, CCA1506, Gate 12 to Beijing, request ATC clearance.

    C: CCA1506, Pudong Delivery, cleared to Beijing, <u>level 9200 meters</u>, PIKAS2X Departure, <u>initial climb and maintain 900 meters</u>, <u>when airborne contact Approach 128.5, squawk 5001</u>.

2.口语练习

答案：

P: Dongfang <u>Delivery</u>, D-YGTV.

C: <u>D-YGTV</u>, Dongfang Delivery.

P: Dongfang Delivery, D-YGTV, <u>Stand</u> 17, <u>destination</u> Guangzhou, <u>Information</u> C, request <u>ATC</u> clearance.

C: <u>D-YGTV</u>, <u>cleared</u> to Guangzhou <u>via flight planned route</u>, ALF11D Departure, initial climb to 600 meters on QNH 1011, maintain 9800 meters on standard, <u>squawk</u> 3475, after departure, <u>contact</u> Approach 128.1.

P: <u>Cleared to Guangzhou via flight planned route, ALF11D Departure, initial climb to 600</u> <u>meters on QNH 1011, maintain 9800 meters on standard, squawk 3475, after departure, contact</u>

Approach 128.1, D-YGTV.

 C: D-YGTV, read-back correct.

3. 口语练习

答案:

(1) **P:** Qingdao Tower, CSN3599, destination Guangzhou with Information C, request ATC clearance.

 C: CSN3599, Information C is valid, cleared to Guangzhou via flight planned route, initial climb to 1200 meters on QNH 1010, XJT-12D Departure, request level change for 8400 meters en route, runway in use 17, squawk 0514. Contact Approach on 123.8 when airborne.

 P: Cleared to Guangzhou via flight planned route, initial climb to 1200 meters on QNH 1010, XJT-12D Departure, request level change for 8400 meters en route, runway in use 17, squawk 0514. Contact Approach on 123.8 when airborne, CSN3599.

 C: CSN3599, read-back correct.

(2) **P:** Hangda Tower, CCA1383, destination WEST with Information X, request ATC clearance.

 C: CCA1383, Information X is valid, cleared to WEST via flight planned route, initial altitude 900 meters on QNH 1009, BRAVO01D Departure, cruising level 8400 meters, runway in use 11, squawk 2230. After departure contact Approach on 123.3.

 P: Cleared to WEST via flight planned route, initial altitude 900 meters on QNH 1009, BRAVO01D Departure, cruising level 8400 meters, runway in use 11, squawk 2230. After departure contact Approach on 123.3, CCA1383.

 C: CCA1383, read-back correct.

2.5 推出开车

1. 完成通话

答案:

(1) CES5357, expect 5 minutes delay due B747 taxiing behind.

(2) PIA852, expect departure 23, start up at own discretion.

(3) SIA 802, push-back and start up approved.

3. 听力练习

录音原文及答案:

(1) **P:** Beijing Ground, CXA505, Stand A6, request start up.

 C: CXA505, start up approved.

(2) **P:** Beijing Ground, CSH100, Stand K15, request start up and push-back. Information M. Our slot-time is 1510 plus 6 minutes.

 C: CSH100, start up and push at time 55.

(3) **P:** Beijing Ground, AFR710 on the apron, request start up and push back.

C: AFR710, expect departure at <u>1310</u>. Start up <u>at own discretion</u> and push back <u>facing east</u>.

（4）**P:** Beijing Ground, JAL410, <u>on Stand D4</u>, request push back.

  C: JAL410, <u>expect 10</u> minutes delay due <u>vehicle breakdown</u>. <u>Stand by Ground 118.9</u>.

4. 思考题

答案：

考虑两种情况：

① 航空器未开始推出：hold position.

② 已开始推出：stop push-back.

2.6　滑出

1. 听力练习

录音原文及答案：

（1）**P:** Pudong ground, CCA981, request taxi.

  C: CCA981, <u>taxi to holding point Runway 35</u> via Taxiway <u>L2, A6, B</u> and <u>A1</u>.

  P: <u>Taxi to holding point Runway 35</u> via Taxiway <u>L2, A6, B</u> and <u>A1</u>, CCA981.

（2）**P:** Ground, ANA211, Stand <u>C8</u>, request taxi to holding point Runway <u>27</u>.

  C: ANA211, taxi to holding point <u>X</u> Runway <u>27</u>.

（3）**P:** <u>Ground</u>, ACA898, Stand D2, request detailed taxi instructions for Runway <u>27</u>.

  C: ACA898, taxi via Taxiway <u>M</u> and <u>Z</u>. Taxi <u>with caution</u>, marked <u>trench right side of</u> taxiway at holding point. Give way to Boeing 737 entering <u>apron</u> from taxiway.

  P: Traffic <u>in sight</u>, ACA898.

（4）**C:** UAL451, hold short of Taxiway <u>N</u>.

  P: Holding short, UAL451.

2. 听说练习

答案：

（1）VY-PPT, go straight ahead at the intersection.

（2）T-CDFG, give way to the aircraft on your left.

（3）D-DJSI, follow the aircraft in front of you.

（4）G-AJGK, there's an aircraft overtaking you on your right.

（5）E-OHIF, taxi straight ahead.

（6）P-FJAB, take the first turning on the right.

（7）D-GHEB, take the first left turn-off.

（8）A-EBQM, take the third turning on the left.

3. 口语练习

答案：

（1）DAL203, give way to an aircraft overtaking you from right.

（2）CXA861, take the first left, report runway vacated.

（3）DLH030, follow the B737 in front of you.

（4）G-ABCD, stop taxi, give way to an aircraft entering the taxiway from the left, he is No.1.

（5）GIA169, stop taxi, the opposite aircraft is going to take the exit on your right.

4. 完成通话

答案：

（1）CCA101, push-back and start up approved, facing east.

CCA101, taxi to CAT I holding point Runway 29 via Taxiway Y, A3, H, A6, A and A7.

（2）CCA102, start up approved.

CCA102, taxi to CAT I holding point Runway 29 via Taxiway H, A6, A and A7.

2.7　起飞

1. 听力练习

录音原文及答案：

（1）**C:** CSN113, <u>are you ready for departure</u>?

　　P: Ready, CSN113.

　　C: CSN113, line up, Runway_09, be ready for immediate departure.

　　P: Lining up, Runway 09, CSN113.

（2）**C:** JAL808, <u>report when ready for departure</u>.

　　P: Ready for departure, JAL808.

　　C: JAL808, <u>behind Boeing 737 on short final line up behind</u>.

　　P: Behind Boeing 737 on final, lining up, JAL808.

　　C: JAL808, correct.

（3）**C:** SIA151, <u>Runway 23, cleared for take-off, report airborne</u>.

　　P: Cleared for take-off, Runway 23, wilco, SIA151.

（4）**C:** CES2014, <u>wind calm, Runway 18, cleared for take-off</u>.

　　P: Cleared for take-off, Runway 18, CES2014.

　　C: CES2014, <u>hold position, cancel take-off clearance, I say again, cancel take-off due dog on runway</u>.

　　P: Holding, CES2014.

（5）**C:** CES5130, <u>wind 320 degrees, 7 m/s, Runway 19, cleared for take-off</u>.

　　P: Taking off, Runway 19, CES5130.

　　C: CES5130, <u>stop immediately, I say again, CES5130, stop immediately</u>.

　　P: Stopping, CES5130.

3. 口语练习

答案：

（1）CCA101, line up, Runway 36.

（2）CCA101, behind landing B737, line up behind, Runway 36.

（3）CCA101, line up and wait, Runway 36.

4. 思考题

答案：

（1）

① 航空器未开始滑跑。

② 航空器已开始滑跑。

（2）

图（a）对应的是①，图（b）对应的是②。

本章综合练习

一、听力练习

录音原文及答案：

P: Dongfang Tower…

C: <u>Station calling Dongfang Tower, say again your call sign.</u>

P: Dongfang Tower, G-JMLF, <u>radio check</u> 118.1.

C: G-JMLF, Dongfang Tower, <u>readability</u> 2, <u>check your transmitter</u> and <u>give me a short count</u>.

P: Roger, G-JMLF.

（A moment later）

P: Dongfang Tower, G-JMLF, 1, 2, 3, 4, 5, how do you read me?

C: <u>G-LF,</u> <u>readability</u> 5.

P: Thank you, G-LF.

P: Dongfang Tower, G-LF, destination Guangzhou, Stand 26, request ATC clearance.

C: G-LF, <u>confirm</u> you have got information C.

P: Negative, G-LF.

C: G-LF, <u>monitor</u> ATIS 128.2, call me when ready.

P: Monitoring 128.2, G-LF.

（A moment later）

P: Dongfang Tower, G-LF, Information C, ready to start.

C: G-LF, Information C is valid, <u>cleared</u> to Guangzhou via flight planned route, GOLFF11D Departure, initial climb to 900 meters on QNH 997, request level change for 10100 meters en route, squawk 5246. <u>Correction</u>, <u>I say again</u>, G-LF, cleared to Guangzhou via flight planned route, GOLFF11D Departure, initial climb to 900 meters on QNH 998, request level change for 10100 meters en route, squawk 5246.

P: Dongfang Tower, G-LF, <u>say again all before</u> request level change for 10100 meters en route.

C: G-LF, GOLFF11D Departure, initial climb to 900 meters on QNH 998.

P: Cleared to Guangzhou via flight planned route, GOLFF11D Departure, initial climb to 900 meters on QNH 998, request level change for 10100 meters en route, squawk 5246, G-LF.

C: ＿G-LF,＿＿read-back correct＿＿.

P: ＿Dongfang Tower,＿＿G-LF＿, request push-back and start up.

C: G-LF, push-back and start up approved.

二、听力练习

录音原文及答案：

C1: <u>CCA981, Pudong Ground, taxi via Taxiway H, Z, N to holding point Runway 35R, hold short of Runway 35R.</u>

P: Taxi via H, Z, N to holding point, Runway 35R, CCA981.

(A moment later)

C1: <u>CCA981, contact Pudong Tower 118.1.</u>

P: Tower 118.1, CCA981.

P: Pudong Tower, CCA981, approaching holding point Runway 35R.

C2: <u>CCA981, hold short of Runway 35R, report B737 on final in sight.</u>

P: Holding short, CCA981.

(A moment later)

P: CCA981, B737 in sight.

C2: <u>CCA981, behind B737 on final, line up behind.</u>

P: Behind B737 on final, line up behind, CCA981.

C2: <u>CCA981, line up.</u>

P: CCA981, lining up.

C2: <u>CCA981, report when ready for departure.</u>

P: CCA981, ready.

C2: <u>CCA981, Runway 35R, cleared for take-off, report airborne.</u>

P: Cleared for take-off, Runway 35R, will report airborne CCA981.

(A moment later)

P: CCA981, airborne 05.

C2: <u>CCA981, contact Approach 118.8, good day.</u>

P: 118.8, good day, CCA981.

三、口语练习

答案：

P: Wuhan Ground, CES521, Destination Chengdu, with Information G, request push-back and start up.

C: <u>CES521, Wuhan Ground, stand by for push-back due B737 taxiing behind.</u>

P: Roger, CES521.

C: <u>CES521, push-back and start up approved, facing north.</u>

(Read-back omitted)

P: Wuhan Ground, CES521, ready for taxi.

C: CES521, taxi to holding point Runway 36R via Taxiway A6 and A.
(Read-back omitted)

C: CES521, taxi with caution, Taxiway C partly covered with snow and ice.

四、听力练习

录音原文及答案：

（1）Taxi straight ahead at the intersection. ——航空器 C

（2）Follow the aircraft in front of you. ——航空器 G

（3）Keep well to the left. There is an aircraft overtaking you on the right. ——航空器 F

（4）Stop taxi. Give way to the aircraft passing you from left to right. ——航空器 D

（5）Stop taxi. Give way to the aircraft entering the taxiway from the left. ——航空器 A

五、故事复述

录音原文：

On the night of Dec. 16, 1997, the crew of Air Canada 646 conducted an ILS approach to Runway 15 at an airport in Canada. The ceiling and visibility were below the minima for instrument approach, but RVR on Runway 15 was 1200 feet, good enough for the approach under Canadian regulations. The captain saw the runway approach lights when the aircraft was 100 feet above decision height. The first officer disconnected the autopilot about 165 feet above ground level and the aircraft began to drift above the glide slope and left of the runway centerline. The first officer reduced thrust to idle. The captain believed the aircraft was not in a position to land safely and commanded a go-around. The aircraft stalled during go-around. The captain and eight passengers were seriously injured.

六、小拓展：监听复诵

参考答案

（1）如果不能及时发现航空器驾驶员的复诵错误，那么航空器将可能偏离期望的管制指令，造成失去间隔或事故的严重后果。

（2）CTL: RJA950(Jordanian950), LA Tower. Via taxiway T cross runway 25R, hold short runway 25L.

　　　PIL: Cross runway 25L, hold short 25R.

（3）指出复诵错误，重复指令，并减慢语速。

第3章　进近管制——离场阶段

3.1　雷达管制用语

2. 听说练习

录音原文及答案：

（1）P: Dongfang Approach, CDG129.

C: CDG129, report heading and level.

P: Heading 135 at 4800 meters, CDG129.

C: CDG129, for identification, turn left heading 020.

P: Turning left heading 020, CDG129.

　（A moment later）

C: CDG129, identified, position 35 kilometers east of airfield.

（2）P: Dongfang Approach, CXA8109, maintaining 5400 meters, estimating HGH 25.

C: CXA8109, Dongfang Approach, report heading.

P: Heading 270, CXA8109.

C: CXA8109, for identification, turn right heading 300.

P: Turning right heading 300, CXA8109.

　（A moment later）

C: CXA8109, not identified, resume own navigation, direct to GYA.

P: Resume own navigation, direct to GYA, CXA8109.

（3）C: CES2308, squawk 3403.

P: Squawking 3403, CES2308.

C: CES2308, radar contact, continue present heading.

P: Continue present heading, CES2308.

（4）C: BAW201, fly heading 330 for spacing, climb to 2400 meters.

P: Heading 330, climbing to 2400 meteres, BAW201.

C: BAW201, continue climb to 3600 meters on standard.

C: BAW201, continue climb to 3600 meters on standard.

C: BAW201, Wuhan Approach.

C: BAW201, radio contact lost, if you read, squawk IDENT.

C: BAW201, squawk observed, will continue radar control.

（5）C: DLH702, Beijing Control.

C: DLH702, Beijing Control.

C: DLH702, reply not received, if you read, turn right heading 120, I say again, turn right heading 120.

C: DLH702, turn observed, resume own navigation, direct JEA, will continue radar control.

（6）C: SAS995, low altitude warning, check your altitude immediately, QNH 1006, the minimum flight altitude is 650 meters.

P: Roger, SAS995.

3. 听力练习

录音原文：

（1）**C:** CCA1427, correction, CCA1247, descend to 3600 meters.

P: Descending to 3600 meters, CCA1247.

（2）**C:** FIN097, turn left heading 120, descend to 2100 meters on QNH 1002.

P: Descending to 1200 meters on QNH 1002, left heading 120, FIN097.

C: Negative, FIN097, descend to 2100 meters, I say again, descend to 2100 meters.

P: Descending to 2100 meters, FIN097.

（3）**C:** CXA8195, confirm squawking 6536.

P1: Affirm, CXA8195.

C: CXA8195, radar contact.

(A moment later)

C: CXA8195, climb to 10100 meters.

P2: Climbing to 10100 meters, CXA8915.

C: Negative, CXA8915, maintain 9500 meters, I say again, CXA8915, maintain 9500 meters, caution similar call sign.

P2: Roger, maintaining 9500 meters, CXA8915.

C: CXA8195, climb to 10100 meters, caution similar call sign.

P1: Climbing to 10100 meters, CXA8195.

（4）**C:** CHH295, turn right heading 170, reduce speed to 250 knots.

P: Speed 170 knots, heading 250, CHH295.

C: Negative, CHH295, turn right heading 170, reduce speed to 250 knots.

P: Heading 170, speed 250 knots, CHH295.

答案：

表 3-1

通话序号	错误的信息	更正（或纠正）后的正确信息
（1）	CCA1427	CCA1247
（2）	descending to 1200 meters	descending to 2100 meters
（3）	CXA8915, climbing to 10100 meters	CXA8195, climb to 10100 meters
（4）	speed 170 knots, heading 250	turn right heading 170, reduce speed to 250 knots

4. 听力练习

录音原文：

（1）**C:** ANA711, position over CU.

P: Roger, ANA711.

（2）**C:** BAW268, resume own navigation, direct YV, magnetic track 030, distance 38 kilometers.

P: Resume own navigation to YV, BAW268.

（3）**C:** AFR105, identified, position 20 kilometers northeast of VYK.

P: Roger, AFR105.

（4）**C:** DLH752, radar service terminated, position 18 kilometers from touchdown, contact

Tower 126.35.

  P: Tower 126.35, DLH752.

（5）**C:** CSN3225, position 40 miles southeast of HUR.

  P: Roger, CSN3225.

答案:

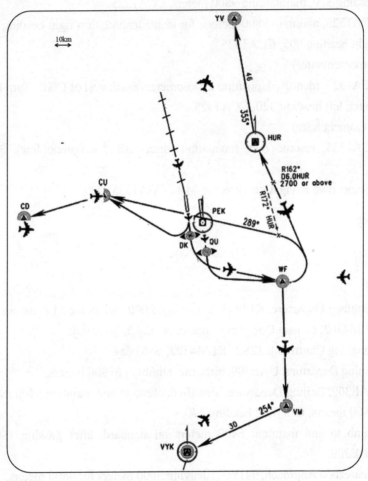

图 3-1

5. 口语练习

答案:

（1）**P:** Dongfang Control, CES2328.

  C: CES2328, Dongfang Control, report heading and level.

  P: Heading 030, maintaining 10700 meters, CES2328.

  C: CES2328, maintain 10700 meters, for identification, turn left heading 345.

  P: Left heading 345, CES2328.

  (A moment later)

P: Dongfang Control, CES2328, on heading 345.

C: CES2328, identified, resume own navigation, direct SHC, position southwest of SHC.

P: Direct to SHC, resume own navigation, CES2328.

（2） **P:** Dongfang Control, CCA1325.

C: CCA1325, Dongfang Control, report heading and level.

P: Heading 320, maintaining 9800 meters, CCA1325.

C: CCA1325, maintain 9800 meters, for identification, turn right heading 005.

P: Right heading 005, CCA1325.

(A moment later)

C: CCA1325, identified, position 20 kilometers northwest of CGK. Turn left heading 320.

P: Roger, left heading 320, CCA1325.

(A moment later)

C: CCA1325, resume own navigation, direct MIL, magnetic track 300, distance 20 kilometers.

P: Resume own navigation, direct to MIL, CCA1325.

3.2 离场指令

2. 听说练习

录音原文：

（1） **P:** Dongfang Departure, KLM4302, leaving 3,000 meters for 5400 meters.

C: KLM4302, contact Dongfang Control on 128.5, good day.

P: Dongfang Control on 128.5, KLM4302, good day.

（2） **P:** Beijing Departure, UAE309, airborne, climbing to 900 meters.

C: UAE309, Beijing Departure, identified, climb to and maintain 5400 meters on standard, after passing 1 800 meters, turn right heading 270.

P: Climb to and maintain 5400 meters on standard, after passing 1800 meters, right heading 270, UAE309.

（3） **P:** Guangzhou Approach, THY021,leaving 3600 meters for 4800 meters.

C: THY021, request further climb from Guangzhou Control on 133.8 at time 1530.

P: 133.8 at 1530, THY021.

3. 听力练习

录音原文及答案：

（1） **P:** Dongfang Approach, CCA4132, airborne at 56, good morning.

C: CCA4132, Dongfang Approach, via CZH 21 Departure, climb to 2100 meters on QNH 1011.

P: Climbing to 2100 meters on QNH 1011, CCA4132.

（2） **P:** Dongfang Departure, UAL971, good afternoon.

C: UAL971, Dongfang Departure, radar contact, climb to and maintain 3600 meters on present heading.

　　　　P: Climbing to 3600 meters on present heading, UAL971.

（3）P: Beijing Departure, ACA083, good morning.

　　　　C: ACA083, Beijing Departure, confirm 80 meters climbing.

　　　　P: Affirm, ACA083.

　　　　C: ACA083, radar contact, turn left heading 030, climb to 2700 meters on QNH 1011 at 2000 ft/min.

　　　　P: Left heading 030, climbing to 2700 meters, QNH 1011, ACA083.

（4）P: Shanghai Departure, JAL4372, good morning.

　　　　C: JAL4372, Shanghai Departure, confirm squawking 4372.

　　　　P: Affirm, JAL4372.

　　　　C: JAL4372, climb straight ahead to 2400 meters, rate of climb not less than 1500 ft/min.

　　　　P: Climbing straight ahead to 2400 meters, JAL4372.

（5）P: Guangzhou Departure, CSN2356, on your frequency.

　　　　C: CSN2356, Guangzhou Departure, radar contact, continue present heading until passing 3000 meters, expedite climb to 1800 meters.

　　　　P: Climbing to 3000 meters, expedite, CSN2356.

4. 听力练习

录音原文：

（1）P: Dongfang Approach, CXA8102, airborne, passing 300 meters climbing to 900 meters.

　　　　C: CXA8102, Dongfang Approach, radar contact, via YCV 01D Departure, climb to 4500 meters on standard, expedite climb until through 3000 meters.

　　　　P: YCV 01D Departure, climbing to 4500 meters on standard, expedite until through 3000 meters, CXA8102.

（2）P: Dongfang Approach, CSN3823, airborne, with you.

　　　　C: CSN3823, Dongfang Approach, radar contact, continue runway heading, climb straight ahead to 4500 meters on standard.

　　　　P: Runway heading, climbing to 4500 meters on standard, CSN3823.

　　　　(A moment later when the aircraft is 30 kilometers north of the airfield.)

　　　　C: CSN3823, resume own navigation direct DPG, position 30 kilometers north of airfield.

　　　　P: Resume own navigation direct DPG, CSN3823.

（3）P: Dongfang Approach, CBJ812, airborne, with you.

　　　　C: CBJ812, Dongfang Approach, identified, via STL 01D Departure, climb to 2100 meters on QNH 1011.

　　　　P: STL 01D Departure, climbing to 2100 meters on QNH 1011, CBJ812.

　　　　(A moment later, when aircraft is over ROM.)

　　　　C: CBJ812 maintain 2100 meters, cancel SID, turn left heading 010 due opposite direction traffic.

P: Maintain level, cancel SID, left heading 010 due traffic, CBJ812.

(A moment later, when aircraft is 10 kilometers left of the route)

C: CBJ812, turn right heading 050, climb to 5100 meters on standard.

P: Right heading 050, climbing to 5100 meters on standard, CBJ812.

(A moment later, when the aircraft is on heading 050 for about 20 kilometers.)

C: CBJ812, resume own navigation direct STL, magnetic track 080, distance 40 kilometers.

P: Resume own navigation direct STL, CBJ812.

选择题答案:

（1）C　　　　（2）B　　　　（3）D

绘图题答案:

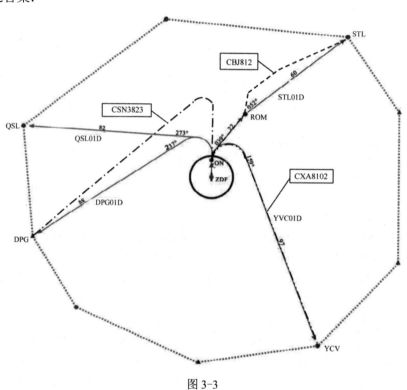

图 3-3

5. 口语练习

答案:

（1）**C:** CES2342, Dongfang Approach, radar contact, follow KG01 Departure, climb to 3000 meters on QNH 1001.

　　P: KG01 Departure, climbing to 3000 meters on QNH 1001, CES2342.

（2）**C:** CSN3443, climb to 4500 meters, contact Dongfang Control 128.35.

　　P: Climbing to 4500 meters, Dongfang Control 128.35, CSN3443.

（3）**C:** CSN3567, continue climb to 2700 meters, turn left heading 330 for spacing.

　　P: Climbing to 2700 meters, left heading 330 CSN3567.

(A moment later)

C: CSN3567, climb to 4500 meters on standard, resume own navigation to ZF, position 40 kilometers southwest of ZF.

P: Climbing to 4500 meters, resume own navigation to ZF, CSN3567.

（4）C: AFR562, Dongfang Approach, identified, follow XS01 Departure, climb to 1800 meters.

P: XS01 Departure, climbing to 1800 meters, AFR562.

(A moment later)

C: AFR562, turn left heading 045 due to opposite direction traffic.

P: Left heading 045, AFR562.

(A moment later)

C: AFR562, turn right heading 110, climb to 4500 meters on standard.

P: Right heading 110, climbing to 4500 meters, AFR562.

(A moment later, the traffic is no factor.)

C: AFR562, climb to and maintain 4500 meters, resume own navigation direct XS, magnetic track 130, distance 30 kilometers.

P: Climbing to and maintain 4500 meters, resume own navigation to XS, AFR562.

(A moment later)

C: AFR562, maintain 4500 meters, contact Dongfang Control on 128.35.

P: Maintain level, Dongfang Control 128.35, AFR562.

3.3　飞行活动通报

2. 听力练习

录音原文及答案：

（1）C: BAW250, Beijing Control, <u>unknown</u> traffic, <u>9 o'clock</u>, 12 kilometers, <u>crossing left to right</u>, Boeing 747,<u> altitude unknown</u>, fasting moving, <u>report traffic in sight</u>.

P: Looking out, BAW250.

(A moment later)

P: Beijing Control, BAW250,<u> negative contact</u>, request <u>vectors</u>.

C: BAW250, turn <u>right</u> heading <u>090 immediately</u>.

P: <u>Right heading 090</u>, BAW250.

C: BAW250, <u>clear of traffic</u>, <u>resume own navigation</u>, <u>direct WTM</u>.

P: Roger, <u>direct WTM,</u> BAW250.

（2）C: THA435, Dongfang Control, <u>unknown traffic</u>, 8 o'clock, 28 kilometers, <u>closing</u>, <u>Airbus A340.</u>

P: Looking out, THA435.

(A moment later)

P: Traffic in sight,<u> looks like a helicopter</u>, <u>well above us</u>, THA435.

C: THA435, <u>clear of traffic</u>.

P: Roger, THA435.

3．口语练习

答案：

（a）CCA1421, traffic 11 o'clock, 15 kilometers, crossing left to right.

（b）CES2006, traffic 1 o'clock, 12 kilometers, crossing right to left.

（c）CSN6719, traffic 9 o'clock, 10 kilometers, south bound, B737, higher level.

（d）B3481, traffic 4 o'clock, same direction, overtaking you on your right.

（e）CHH7378, traffic 10 o'clock, 15 kilometers, northwest bound, A320, level below you.

（f）AAR334, opposite direction traffic, 12 o'clock, 50 kilometers, climbing through your level.

4．口语练习

答案：

C: HDA623,Shanghai Approach, traffic, 1 o'clock, 15 kilometers, crossing right to left, report traffic in sight.

P: Looking out, HDA623.

P: Shanghai Approach, HDA623, negative contact, request vectors.

C: HDA623, turn left heading 330 to avoid traffic.

P: Left heading 330, HDA623.

(A moment later)

C: HDA623, clear of traffic.

P: Roger, HDA623.

5．听说练习

录音原文：

（1）**Scenario:** Your call sign is AFR198. You are maintaining 2100 meters on heading 030.
　　C: AFR198, traffic, 11 o'clock, same direction, 1800 meters.

（2）**Scenario:** You are on heading 150 at an altitude of 3000 meters. Your call sign is HDA934.
　　C: HDA934, traffic, 1 o'clock, 20 kilometers, 737, descending through your altitude.

（3）**Scenario:** Your call sign is BAW728. You just leveled off at 1200 meters on a heading of 295.
　　C: BAW728, traffic, 2 o'clock, westbound, Mode C indicating 1500 meters, not verified.

（4）**Scenario:** Your call sign is FIN912, you just vacated 4800 meters descending to 2 700 meters.
　　C: FIN912, unknown traffic, 2 o'clock, 12 kilometers, southwest bound, fast moving.

（5）**Scenario:** Your call sign is ACA943. You are climbing from 3600 meters to 4500 meters on a heading of 070.
　　C: ACA943, traffic, 1 o'clock, 15 kilometers, climbing through 4200 meters.

答案：

（1）The traffic is 300 meters below me, on my left.

（2）The traffic is descending through my level, on my right.

（3）The traffic is 300 meters above me, on my right.

（4）The traffic is on my right, level unknown.

（5）The traffic is 600 meters above me, on my right.

本章综合练习

一、完成通话

录音原文及答案：

（1）**C:** AMU525, <u>radar</u> contact, 12 miles west of LKO, <u>continue</u> present <u>heading</u> until crossing WHA.

（2）**C:** CSS467, <u>turn</u> right heading 070 <u>immediately</u> to avoid unidentified <u>traffic</u>, 10 o'clock, 8 miles, height unknown.

（3）**C:** HDA525, <u>radar</u> service terminated, <u>resume</u> own navigation <u>direct</u> SHR.

（4）**C:** CPA555, <u>squawk</u> 4562.

　　P: 4562, CPA555.

　　C: CPA555, <u>confirm</u> squawk.

　　P: Squawking 4562,CPA555.

　　C: CPA555, <u>reset</u> 4562.

　　P: Resetting 4562, CPA555.

　　C: CPA555, <u>radar contact</u>. Climb to 4,500 meters.

（5）**C:** CSN3369, <u>unknown</u> traffic, 10 o'clock, 10 kilometers, <u>crossing</u> left to right.

　　P: Negative contact, CSN3369. <u>Request vectors</u>.

（6）**C:** EVA712, wind 340 degrees, 2 m/s, Runway 36, <u>cleared for take-off</u>.

　　P: Cleared for take-off, Runway 36, EVA712.

　　　（A moment later）

　　P: Departure, EVA712, airborne at 15, passing 100 meters climbing to 900 meters.

　　C: EVA712, <u>radar contact</u>, <u>continue climb</u> to 3000 meters, report <u>passing</u> 1800 meters. <u>Cancel</u> SID, <u>turn left heading</u> 330.

　　P: Climbing to 3000 meters, left heading 330, wilco, EVA712.

　　　（A moment later）

　　P: EVA712, passing 1800 meters.

　　C: EVA712, roger, report reaching 3000 meters.

　　P: Wilco, EVA712.

　　　（A moment later）

　　P: EVA712, approaching 3000 meters.

　　C: EVA712, <u>resume own navigation</u> direct JTG, <u>position</u> 60 kilometers northwest of the field.

　　P: Direct JTG, EVA712.

　　C: EVA712, climb to 5400 meters on standard, <u>climb at</u> 2000 ft/min <u>or greater</u>.

P: Expediting climb to 5400 meters, EVA712.

(A moment later)

P: EVA712, maintaining 5400 meters.

C: EVA712, <u>contact</u> Dongfang Control on 120.1, good day.

P: Dongfang Control on 120.1, good day, EVA712.

二、听力练习

答案:

(1) Radar contact.

(2) Turn right heading 130.

(3) Continue runway heading.

(4) Continue present heading until 1800 meters.

(5) Squawk 5327 and IDENT.

(6) Climb straight ahead.

(7) Follow CHEDY One RNAV Departure.

(8) Contact Dongfang Control on 120.1.

(9) Proceed direct to LXA.

(10) Radio contact lost, if you read, turn right heading 030.

(11) Resume own navigation, position 25 kilometers southwest of LJA.

(12) Opposite direction traffic, crossing left to right, fast moving.

(13) Negative contact, request vectors.

(14) Same direction traffic overtaking you on the left.

(15) Clear of traffic.

三、听力练习

录音原文及答案:

(1) **C:** <u>UAL058, squawk 4261.</u>

   P: Squawking 4261, UAL058.

   C: <u>UAL058, confirm squawk.</u>

   P: Squawking 4261, UAL058.

   C: <u>UAL058, radar contact, turn right heading 150, vectoring for spacing, climb to and maintain 2400 meters, expect to resume LADIX One RNAV Departure.</u>

   P: Right heading 150, climbing to 2400 meters, LADIX One RNAV Departure, UAL058.

   C: <u>UAL058, climb to 4500 meters, cleared direct AA011, resume LADIX One RNAV Departure, comply with restrictions.</u>

   P: Climbing to 5400 meters, direct AA011, LADIX One RNAV Departure, wilco, UAL058.

   C: <u>UAL058, negative, climb to 4500 meters.</u>

   P: Climbing to 4500 meters, UAL058.

（2）**C:** MGL101, Zhuhai Approach, avoiding action, turn right immediately heading 090, unknown traffic at 12 o'clock.

　　　P: Right heading 090, MGL101.

（3）**C:** MAS802, Xi'an Approach, unknown traffic at 8 o'clock.

　　　P: Looking out, MAS802.

　　　P: Xi'an Approach, MAS802, traffic in sight.

　　　C: MAS802, now clear of traffic.

　　　P: Roger, MAS802.

（4）**C:** HVN1302, Guangzhou Departure, unknown traffic, 10 o'clock, 8 miles, crossing left to right, report traffic in sight.

　　　P: Negative contact, request vectors, HVN1302.

　　　C: HVN1302, turn left heading 050.

　　　P: Left heading 050, HVN1302.

　　　C: HVN1302, clear of traffic, resume own navigation to LMN, magnetic track 070, distance 27 miles.

　　　P: Roger, track 070, distance 27 miles, HVN1302.

（5）**P:** Dongfang Approach, DKH2371,passing the zone boundary.

　　　C: DKH2371, contact Dongfang Information on 125.75 for flight information.

　　　P: 125.75, DKH2371.

（6）**P:** Dongfang Information, G-ABCD, Yun-7, departed from Dongfang airport 1332Z, 1800 meters, VFR to Nanping airport, estimating Nanping airport 1442Z, request Nanping airport weather.

　　　FIS: G-ABCD, Dongfang Information, Nanping airport not available, call Dongfang Control on 128.7.

选择题答案：

（1）D　　　（2）A　　　（3）B　　　（4）C

四、故事复述

录音原文：

At a large international airport a Boeing 737 and a Cessna 210 were both cleared to line up and wait on different but intersecting runways. The 737 was then cleared for take-off, which the crew acknowledged. Unfortunately, the Cessna pilot, expecting to hear "cleared for take-off" for himself, started his take-off without fully comprehending the transmission. The controller did not pick up on a duplicated acknowledgement from two different aircraft. The two aircraft met at the intersection, and only a dangerous early rotation by the 737 pilot enabled both aircraft to escape a collision. The miss distance was only thirty meters.

五、小拓展：鸟击

(1) bird strike; flock of birds; windshield; cracked; burn off; persons on board (POB); fuel on board

(2) Delta 2292 passing 2800 feet, we're climbing now to 5000, but we just hit something. We're not sure what it was. And 3000 to 5000 and we're declaring an emergency and we'll get right back with you.

We hit something. We don't know what we hit but it was a huge hit at about—we don't know—it was about 1500 feet. It was right through—kinda over the stadium making that left turn on the way doin' climb. It was a really hard hit on the airplane. Maybe just a big bird or something but we don't know what kind of damage we have.

(3) We need to go somewhere else. We're a little heavy for that but the engines are working good and everything looks pretty good, but we definitely hit something so we're not continuing this flight. We're going back to JFK. So, could you give us somewhere—just some vectors for a few minutes until we get this sorted out?

We're actually gonna need just a little bit to work out something so if you just set us somewhere where we can hold for just a little bit.

(4) Do you wanna turn back to LaGuardia or do you wanna go somewhere else?

Do you wanna go higher for a while or does 5000 work out for you?

(5) When you have a chance, I'm just gonna need the fuel and souls on board.

When able, would you just tell me the amount of fuel you have remaining in pounds.

(6) Whenever you're able, I'll start to keep asking you questions. They just wanna know the nature of the emergency.

第4章　区　域　管　制

4.1　高度信息

1. 词汇练习

答案：

（1）through, to

（2）at

（3）from

（4）back

（5）away

（6）on

（7）over

（8）right

2. 口语练习

答案：

Student A

（1）ALK866, descend to 4500 meters at 2000 ft/min or greater.

（2）UAL543, turn left heading 120, descend to 5400 meters.

（3）CCA891, descend to 6300 meters, contact Approach on 120.4.

（4）AFL209, after passing ANRAT, descend to 6600 meters.

（5）ABW297, descend to 7200 meters, expedite until passing 8400 meters.

Student B

（6）AZA98, climb to 8900 meters, expedite until passing 7500 meters.

（7）AIC356, climb to reach 9200 meters by VYK, report reaching.

（8）HVN4648, turn right heading 130, descend to and maintain 6900 meters.

（9）AFL65, the preceding traffic reported turbulence at 9200 meters. Can you accept 10400 meters?

（10）JAL397, descend to reach 7200 meters at time 05.

3. 完成通话

答案：

C: CCA1331, climb to 8900 meters, report reaching.

P: Climbing to 8900 meters, CCA1331.

P: CCA1331, approaching 8 900 meters.

C: CCA1331, continue climb to 10100 meters.

P: Climbing to 10100 meters. Is 11300 meters available? CCA1331.

C: CCA1331, negative, 11300 meters is unavailable. Can you accept 10700 meters?

P: Negative.

C: CCA1331, contact Shanghai Control 124.1 .

P: 124.1 for Shanghai Control, CCA1331.

4. 听力练习

录音原文及答案：

（1）**C1:** CCA1527, contact Dongfang Control on 120.1, good bye.

　　P: 120.1, CCA1527, good bye.

　　　（A moment later）

　　P: Dongfang Control, CCA1527, good morning.

　　C2: CCA1527, Dongfang Control, good morning.

　　P: Climbing to 4500 meters, heading 260, CCA1527.

　　C2: CCA1527, continue climb to 6300 meters, report reaching, continue present heading.

　　P: Climbing to 6300 meters, maintaining present heading, CCA1527.

　　　（A moment later）

　　P: CCA1527, reaching 6300 meters.

　　C2: Roger, CCA1527, climb to 7500 meters.

　　P: Climbing to 7500 meters, CCA1527.

　　C2: CCA1527, contact Beijing Control 129.4, good day.

P: 129.4, CCA1527, good day.

（2） **C:** <u>AFR116, squawk IDENT</u>.

P: Squawking IDENT, AFR116.

C: <u>AFR116, radar contact, descend to 7800 meters</u>.

P: Leaving 9800 meters descending to 7800 meters, AFR116.

C: <u>AFR116, continue descent to 6000 meters, report passing 6600 meters</u>.

P: Descending to 6000 meters, wilco, AFR116.

P: AFR116, passing 6600 meters, descending.

C: <u>AFR116, confirm leaving level 6600 meters</u>.

P: Affirm, AFR116.

（3） **C1:** <u>ACA018, contact Beijing Control on 126.5</u>.

P: 126.5, good day, ACA018.

（A moment later）

P: Beijing Control, ACA018, good morning.

C2: <u>ACA018, Beijing Control, good morning</u>.

P: ACA018, 4800 meters maintaining, heading 230.

C2: <u>ACA018, climb to 9200 meters, expedite until passing 7200 meters</u>.

P: Climbing to 9200 meters, expedite until passing 7200 meters, ACA018.

C2: <u>ACA018, turn right, correction, turn left heading 180</u>.

P: Left heading 180, ACA018.

C2: <u>ACA018, contact Shanghai Control on 128.9</u>.

P: 128.9, good day, ACA018.

（4） **C:** <u>UAL889, climb to 8100 meters</u>.

P: Climbing to 8100 meters, UAL889. Is 8900 meters available?

C: <u>UAL889, 8900 meters is unavailable due to heavy traffic. Can you accept 9500 meters</u>?

P: Negative, UAL889.

C: <u>UAL889, climb then to 8100 meters, report reaching</u>.

P: Climbing to 8100 meters, UAL889.

（A moment later）

P: UAL889, reaching 8100 meters.

C: <u>UAL889, contact Dongfang Control on 128.9, good day</u>.

P: Dongfang Control on 128.9, good day, UAL889.

（5） **P:** Dongfang Control, CES2020, over TER, maintaining 9800 meters, with you.

C: <u>CES2020, Dongfang Control, radar contact, descend to 8400 meters</u>.

P: Leaving 9800 meters, descend to 8400 meters, CES2020.

C: <u>CES2020, continue descend to 5400 meters, report passing 7200 meters</u>.

P: Descending to 5400 meters, CES2020.

（A moment later）

P: CES2020, passing 7200 meters.

C: <u>CES2020, confirm passing 7200 meters.</u>
P: Affirm, CES2020.

4.2　位置信息

1. 听力练习

录音原文：

（1）Dongfang Control, CCA1212, over YQG, at 15, maintaining 8400 meters, estimating BTO at 35, next VSK.

（2）Dongfang Control, CSN6532, over HXN, at 23, maintaining 9200 meters, estimating ADE at 45, next WSI.

（3）Dongfang Control, CES5245, over CFJ, at 34, maintaining 10100 meters, estimating LMP at 54.

（4）Donfang Control, CDG3387, over NRT, at 50, maintaining 9500 meters, estimating ZUI at 1120.

（5）Dongfang Control, CHH7849, over CFJ, at 10, maintaining 8900 meters, estimating NRT at 35.

答案：

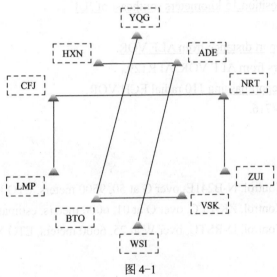

图 4-1

2. 完成通话

答案：

（1）Report passing VICIS.
（2）Omit position reports until DELTA.
（3）Report passing 210 radial ROM VOR.
（4）Report 15 miles from MKE DME.
（5）Resume position reporting.
（6）Position 18 kilometers from touchdown.
（7）Position 38 kilometers northeast of the field.
（8）Position 35 kilometers east of A593.

（9）Magnetic track 030, distance 12 kilometers.

（10）CSN1232, now passing the field.

3. 听力练习

录音原文及答案：

（1）**C:** CES5115, report passing ECH VOR.

　　P: ECH at 35, maintaining 9200 meters, estimating WISKY at 53, CES5115.

　　C: CES5115, roger.

（2）**C:** HDA306, report 35 kilometers from MKE DME.

　　P: Wilco, HDA306.

（3）**C:** PAL331, omit position reports until COAST.

　　P: Roger, PAL331.

　　　（A moment later）

　　C: PAL331, resume position reporting.

　　P: Roger, PAL331.

（4）**C:** ACA085, position 15 kilometers southeast of JUI.

　　P: Roger, ACA085.

（5）**C:** AFR129, report distance from ALF VOR.

　　P: 47 kilometers from ALF VOR, AFR129.

（6）**C:** BAW716, report passing 110 radial ECH VOR.

　　P: Roger, BAW716.

4. 完成通话

答案：

（1）**P:** Dongfang Control, N-B241E, over G at 50, 9500 meters, ETO H at 0215.

（2）**P:** Dongfang Control, ZK-ELC, over O at 01, 6000 meters, estimating P at 55.

（3）**P:** Dongfang Control, D-RSTU, over W at 25, 6600 meters, ETO X at 1720.

4.3　航路等待

2. 口语练习

答案：

AFR182, cleared to HO, maintain 7000 feet, hold inbound track 011 degrees, right hand pattern, outbound time one minute, expect further clearance at 20.

3. 听力练习

原音原文：

（1）**P:** Dongfang Control, CDG4951, request detailed holding instructions.

　　C: CDG4951, proceed to KIG, maintain 7200 meters, hold inbound track 020 degrees, left hand pattern, outbound time one and a half minutes, expect further clearance at 19.

（2）**P:** Dongfang Control, AAL289, request detailed holding instructions.

　　C: AAL289, cleared to the 120 radial of the KIG VOR at 45 kilometers DME Fix. Maintain 6900 meters, hold east, left hand pattern, outbound time one and a half minutes, expect further clearance at 36, holding speed at or below 550 kilometer per hour.

（3）**P:** Dongfang Control, CPA636, request detailed holding instructions.

　　C: CPA636, cleared to the 120 radial of the KIG VOR at 30 kilometers DME Fix. Maintain 8900 meters, hold between 25 kilometers and 35 kilometers DME, right hand pattern, expect further clearance at 47.

答案：

4.4　RVSM 运行

1. 完成通话

答案：

（1）CCA1324, confirm RVSM approved.

（2）N-898EF, unable issue clearance into RVSM airspace, maintain 8400 meters.

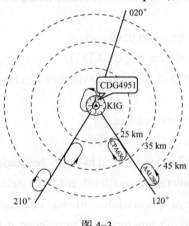

图 4-3

2. 完成通话

答案：

C: CBJ5323, advise if able to proceed parallel offset.

P: Affirm, CBJ5323.

C: CBJ5323, proceed offset 2 miles right of A463.

P: Offset 2 miles right of A463, CBJ5323.

4.5　绕飞雷雨

1. 完成通话

答案：

（1）CSN6721, right turn approved, report clear of weather.

（2）LKE9837, the thunderstorm is rapidly approaching you.Turn right 20 degrees, track out 30 kilometers to circumnavigate it.

（3）GCR7891, turn right to detour the thunderstorm.

（4）CCA1310, previous aircraft reported light turbulence at all adjacent levels of 8100 meters.

（5）CES2831, keep us advised if conditions continue or get worse (deteriorate).

2. 听力练习

录音原文：

C1: CCA1331, contact Dongfang Control on 120.8, good day.

P1: 120.8, good day.

P1: Dongfang Control, CCA1331, over KG, maintaining 8100 meters, request a higher level.

C2: CCA1331, Dongfang Control, radar contact, maintain 8100 meters, will call you back for further climb.

P1: Maintaining 8100 meters, CCA1331.

C2: CCA1527, descend to and maintain 6000 meters, expedite descent until passing 7800 meters.

P2: Descend to and maintain 6000 meters, expedite passing 7800 meters, CCA1527.

　　(A moment later)

C2: CCA1331, climb to and maintain 9500 meters.

P1: Climb to and maintain 9500 meters, CCA1331.

P1: Dongfang Control, CCA1331, request 20 degrees heading change right of track to avoid build-ups.

C2: Roger, CCA1331, report new heading.

P1: Heading 250, CCA1331.

C2: CCA1331, how many miles do you need to be on the heading 250?

P1: CCA1331, about 20 kilometers, then I'll offset 15 kilometers east of A493.

C2: CCA1331, heading change approved. Report clear of weather.

P3: Dongfang Control, CSN6903, any reports of turbulence at 10400 meters?

C2: CSN6903, previous aircraft reported turbulence only up to 10100 meters.

P3: CSN6903, request climb to 10400 meters.

C2: CSN6903, climb to and maintain 10400 meters.

P3: Climbing to 10400 meters, CSN6903.

答案：

（1）A　　（2）B　　（3）B　　（4）C　　（5）B

4.6　航空器加入、穿越或离开航路

听力练习

录音原文及答案

（1）**P:** Dongfang Control, C-GNDK, request airways crossing V22 at KOL.

C: C-GNDK, Dongfang Control.

P: C-GNDK, MA60, passing JUI heading 290, maintaining 1800 meters, VMC, ETO KOL at 1115.

C: C-GNDK, cross V22 at KOL not before 1115.

P: Cross V22 at KOL not before 1115 at 1,800 meters, C-GNDK.

（2）**P:** Dongfang Control, N550SD, request joining clearance at GOF.

C: N550SD, Dongfang Control.

P: N550SD, Cessna 310, 19 miles west of GOF intersection, heading 090, maintaining 2400 meters, outbound from HTE VOR/DME, ETO GOF at 45, B2 for Beian. Request 3000 meters.

C: N550SD, join B2 at GOF intersection, climb and maintain 3000 meters, maintain own separation and contact Beian Tower on 122.1.

本章综合练习

一、听说练习

答案：

（1）Climb to 5400 meters immediately.

（2）Turn left heading 230.

（3）Go around, I say again, go around. Traffic on 34 Right.

（4）Climb to 9200 meters, expedite until passing 7200 meters.

（5）Descend to 2400 meters, report passing 3600 meters.

二、听力练习

录音原文：

指令 1

C: N-3421E, Dongfang Control.

P: Dongfang Control, N-3421E.

C: N-3421E, descend to FL 310 to be level by KIG. Hold at KIG FL 310. Inbound track 322 degrees, right hand pattern, report at KIG.

指令 2

C: ZK-DAE, Dongfang Control.

P: Dongfang Control, ZK-DAE.

C: ZK-DAE, hold inbound on the 249 radial MKE VOR between 15 miles and 25 miles DME, right hand pattern.

答案：

（1）Descent;

（2）FL 310;

（3）Inbound track 322 degrees;

（4）Inbound on the 249 radial MKE VOR;

（5）Between 15 miles and 25 miles MKE DME.

三、完成通话

录音原文及答案：

C1: <u>CPA7402, line up and wait.</u>

P: Line up, CPA7402.

C1: <u>CPA7402, wind calm, Runway 18, cleared for take-off.</u>

P: Cleared for take-off, Runway 18, CPA7402.

C1: <u>CPA7402, contact Dongfang Departure 119.7.</u>

P: 119.7, CPA7402.

P: Dongfang Departure, CPA7402, climbing to 2100 meters, passing 1200 meters.

C2: <u>CPA7402, Dongfang Departure, identified, climb to 6300 meters.</u>

P: Climbing to 6300 meters, CPA7402.

C2: <u>CPA7402, contact Dongfang Control 134.7.</u>

P: 134.7, CPA7402.

P: Dongfang Control, CPA7402, climbing to 6300 meters.

C3: <u>CPA7402, Dongfang Control, climb to 9500 meters.</u>

P: Climbing to 9500 meters, CPA7402. Any reports of turbulence at 9500 meters this morning?

C3: <u>CPA7402, only up to 9200 meters. However there have been others at 9500 meters who haven't reported any.</u>

P: Roger, thanks, CPA7402.

四、完成通话

录音原文及答案：

C0: <u>CCA 1331, contact Dongfang Control on 120.8, good day.</u>

P: Dongfang Control 120.8, CCA1331, good day.

　　(A moment later)

P: Dongfang Control, CCA1331, good morning.

C1: <u>CCA1331, Dongfang Control.</u>

P: CCA1331, over HTO, maintaining 4500 meters.

C1: <u>CCA1331, radar contact, climb to 6300 meters.</u>

P: Climbing to 6300 meters, CCA1331.

　　(A moment later)

P: CCA1331, maintaining 6300 meters.

C1: <u>CCA1331, climb to 8900 meters, expedite climb until passing 7500 meters.</u>

P: Climbing to 8900 meters, expediting to 7500 meters, CCA1331.

　　(A moment later)

P: CCA1331, approaching 8900 meters.

C1: CCA1331, roger.

(A moment later)

C1: <u>CCA1331, contact Beijing Control on 124.3, good day</u>.

P: Beijing Control 124.3, CCA1331, good day.

(A moment later)

P: Beijing Control, CCA1331, good morning.

C2: <u>CCA1331, Beijing Control</u>.

P: CCA1331, passing GOL 03, maintaining 8900 meters, estimating INK at 18.

C2: <u>CCA1331, roger</u>.

(A moment later)

P: CCA1331, request climb to 10100 meters due to moderate air turbulence.

C2: <u>CCA1331, 10100 meters is unavailable at this moment due to heavy traffic, can you accept 10700 meters</u>?

P: Affirm, CCA1331.

C2: <u>CCA1331, climb to 10700 meters</u>.

P: Climbing to 10700 meters, CCA1331.

(A moment later)

P: CCA1331, maintaining 10700 meters.

C2: <u>CCA1331, contact Shanghai Control on 131.1, good day</u>.

P: 131.1 for Shanghai, CCA1331, good day.

(A moment later)

P: Shanghai Control, CCA1331, over MKE, maintaining 10700 meters.

C3: <u>CCA1331, Shanghai Control, radar contact</u>.

(A moment later)

C3: CCA1331, <u>descend to 8400 meters, rate of descent not less than 2000 ft/min</u>.

P: Descending to 8400 meters, not less than 2000, CCA1331.

(A moment later)

C3: <u>CCA1331, continue descent to 6000 meters, report reaching</u>.

P: Descending to 6000 meters, will report reaching, CCA1331.

(A moment later)

P: CCA1331, maintaining 6000 meters.

C3: <u>CCA1331, contact Shanghai Arrival on 119.3, good day</u>.

P: 119.3 for Shanghai Arrival, CCA1331, good day.

五、听说练习

录音原文:

P1: Dongfang Control, CES2328, passing FYG 37, maintaining 11000 meters, estimating MIDOX 43.

C: CES2328, Dongfang Control, roger, maintain 11000 meters, next report at LKO.

P1: Maintain level, wilco, CES2328.

P2: Dongfang Control, CDG328, maintaining 10400 meters, ETO OBLIK 45.

C: CDG328, Dongfang Control, radar contact, after OBLIK, turn right heading 230 for spacing.

P2: After OBLIK, right heading 230, CDG328.

C: CXA8459, turn left heading 300 immediately due to crossing traffic. Break, break, CSZ9886, turn left heading 090, climb to 9800 meters immediately.

P3: Left heading 300, CXA8459.

P4: Left heading 090, climbing to 9800 meters, CSZ9886.

(CDG328 has traveled 10 miles on heading 230.)

C: CDG328, resume own navigation, direct to BEIWU, position 10 miles northwest of BEIWU.

P2: Roger, resume own navigation to BEIWU, CDG328.

(CXA8459 and CSZ9886 are clear when they are on the present heading for about 10 kilometers.)

C: CSZ9886, clear of traffic, maintain 9800 meters, resume own navigation direct ZF, magnetic track175, distance 15 miles.

P4: Maintain level, resume own navigation to ZF, CSZ9886.

C: CXA8459, clear of traffic, resume own navigation to ZF, magnetic track 070, distance 20 miles.

P3: Roger, resume own navigation to ZF, CXA8459.

问答题答案:

（1）CES2328 flies over MIDOX at 43.

（2）The controller instructs CXA8459 to make a left turn (heading 300) and issues heading and level change instructions for CSZ9886, which is "turn left heading 090, climb to 9800 meters immediately".

（3）CXA8459, clear of traffic, resume own navigation to ZF, magnetic track 070, distance 20 miles.

绘图题答案:

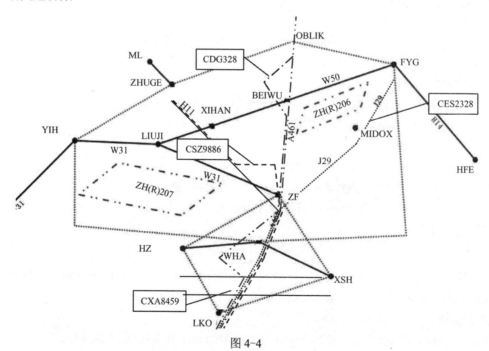

图 4-4

六、故事复述

录音原文：

On November 24, 1971, a man calling himself Dan Cooper boarded a Northwest Airlines Boeing 727 bound for Seattle Washington. Soon after takeoff, Cooper stated he had a bomb. He demanded $200,000 in cash and several parachutes. After the plane landed at Seattle, the passengers were allowed to leave. Cooper and four crew members took off with his instructions to fly towards Mexico. The pilot was instructed to fly no higher than 10000 feet and below 200 mph. He asked the flight attendant how to open the tail stairway and ordered her to the front of the plane. Shortly after, the crew felt a thud and Cooper jumped from the plane with a package of money tied to his waist. He was never heard from again. Despite a massive search, no sign of him was ever found.

七、小拓展：空中失火

（1）fire warning; landing priority; evacuate; dangerous goods; engine; cargo hold; cabin; lavatory; galley oven; wheel well

（2）第一次：JBU2401 is declaring an emergency at this time. We'd like to return to Kennedy. We have some type of electrical burning smell up here. Souls on board 155 and roughly about 4 hours of fuel on board.

第二次：There is no smoke. There was a strong electrical smell and right now we're just trying to run a couple of checklists.

第三次：There was no visible smoke, however we had a very strong odor what seems to be an electrical fire or burning smell.

第四次：We have no indication of a fire and we have no visible smoke, however, it was a very strong odor. That's why we wanna return to the airport.

（3）Understand it's a burning smell. Is that correct?

（4）Is there anything you need at this time?

I just wanna make sure there is nothing else that you need at this time.

（5）Do you have souls on board currently? Can you give me a fuel remaining in pounds?

（6）

Endeavor3696：We got an emergency coming in from the south. I'm gonna change you back to — it's gonna be an ILS 31 R approach for you based on that emergency.

Delta 405:We have an emergency and he is to land on the left and you're gonna stay on the right.

Delta 1：

I need you to keep your speed up. I got an emergency that's gonna be coming in for 31L so keep your speed up as long as possible.

第5章 进近管制——进场阶段

5.1 进场及进近

1. 听力练习

录音原文：

（1）Dialogue 1

    P: Beijing Approach, SIA802, 5100 meters, over VYK with you.

    C: SIA802, Beijing Approach, VYK-03 arrival, expect Runway 01, Information V is valid.

    P: VYK-03 Arrival, Runway 01, SIA802.

    C: SIA802, descend to and maintain 3000 meters on QNH 1011.

    P: Descending to 3000 meters on QNH 1011, SIA802.

    C: SIA802, reduce speed to 250 knots.

    P: Reducing speed to 250 knots, SIA802.

        (A moment later)

    C: SIA802, contact Final East on 119.0, good day.

    P: 119.0, SIA802, good day.

（2）Dialogue 2

    P: Beijing Final, SIA802, descending to 3000 meters, on VYK-03 arrival with you.

    C: SIA802, Beijing Final, descend to and maintain 900 meters, reduce speed to 220 knots, Runway 01.

    P: Descending to 900 meters, reduce speed to 220 knots, SIA802.

    C: SIA802, turn left heading 030, radar vector for ILS Runway 01.

    P: Left heading 030, SIA802.

        (A moment later)

    C: SIA802, turn left heading 270 for base.

    P: Turning left heading 270, SIA802.

        (A moment later)

    C: SIA802, traffic at two o'clock, 9 kilometers, 1800 meters northbound, is approaching for Runway 36L.

    P: Copy the traffic, SIA802.

    C: SIA802, turn right heading 340, intercept localizer Runway 01.

    P: Turning right heading 340, intercept localizer Runway 01, SIA802.

        (A moment later)

    P: SIA802, established on localizer Runway 01.

    C: SIA802, cleared for ILS approach Runway 01.

    P: Cleared for ILS approach Runway 01, SIA802.

        (A moment later)

P: SIA802, established ILS Runway 01.

C: SIA802, contact Tower on 118.6, good day.

P: Contact Tower on 118.6, SIA802, good day.

答案：

（1）VYK-03 arrival；V；3000 meters；1011；250 knots；119.0.

（2）220 knots；270；two；9 kilometers；36 L；340；01；118.6.

2. 口语练习

答案：

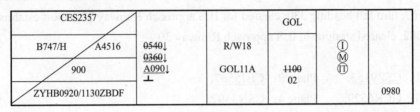

图 5-2

3. 完成通话

录音原文及答案：

P: Dongfang Approach, CES5129, maintaining 5400 meters, Information M.

C: CES5129, Dongfang Approach, radar contact, expect ILS approach, Runway 36, Information M is valid.

P: ILS approach, Runway 36, CES5129.

C: CES5129, descend to 2400 meters on QNH 1011, report reaching.

P: Descending to 2400 meters on QNH 1011, wilco, CES5129.

(A moment later)

P: CES5129, maintaining 2400 meters.

C: CES5129, descend to 900 meters.

P: Descending to 900 meters, CES5129.

(A moment later)

C: CES5129, turn right heading 030, cleared for ILS approach, Runway 36, report established localizer.

P: Right heading 030, cleared for ILS approach, Runway 36, will report established localizer, CES5129.

(A moment later)

P: CES5129, established localizer.

C: CES5129, position 18 kilometers from touchdown, radar service terminated, contact Tower on 118.1, good day.

P: Contact Tower on 118.1, good day, CES5129.

4. 听说练习

（1）

录音原文：

GCR6529, continue present heading , extend downwind.

CSN3392, maintain present speed, light wind shear reported by preceding aircraft at downwind.

CSZ9153, position 10 kilometers from touchdown, radar service terminated, contact tower on 118.7, good day.

EVA780, proceed direct to CGO, expect turn right to join the downwind in 4 minutes.

CBJ5567, turn left heading 330, cleared for ILS approach Runway 30, report established.

AMU082, cleared straight-in ILS approach Runway 30.

答案：

Plane 1：CSZ9153,   Plane 2：CBJ5567,

Plane 3：GCR6529,   Plane 4：CSN3392,

Plane 5：AMU082,   Plane 6：EVA780.

（2）

答案：

① C

② E

③ B

④ A

⑤ F

⑥ D

本章综合练习

一、听力练习

录音原文：

（1）CQH8876, climb to and maintain 4500 meters on standard.

（2）CQH8867, radar contact, vectoring for ILS approach, Runway 36.

（3）CES5263, turn right heading 330, cleared for ILS approach, Runway 36, report established.

（4）CES5623, cancel ALF Arrival, turn right heading 360.

（5）CSH9260, reduce speed to 180 knots, turn right heading 270.

（6）CSH9206, identified, position 18 kilometers northeast of the field.

（7）UAL835, descend to 3000 meters at 2000 ft/min.

（8）UAL858, radar service terminated, contact Tower 118.1, good day.

（9）Dongfang Approach, CSN3554, over CH, 5100 meters maintaining, Information S.

（10）Dongfang Approach, CSN3524, passing 100 meters, climbing to 900 meters on QNH 1003.

答案：

① CQH8867;   ② CSN3524;   ③ CSH9206;   ④ UAL858;   ⑤ CQH8876;

⑥ UAL835;   ⑦ CSH9260;   ⑧ CES5263;   ⑨ CSN3554;   ⑩ CES5623.

二、完成通话

答案:

(1) CCA124, continue approach, position 20 kilometers from touchdown, radar service terminated, contact Tower on 118.6, good day.

(2) CSN378, reduce speed to 180 knots.

(3) CCA1312, descend to and maintain 900 meters, cleared for ILS approach, Runway 36L, report established localizer.

三、听说练习

1.

录音原文:

(1) C: D-GIWL, Dongfang Appoach, radar contact, climb to 1200 meters.

P: Climbing to 1200 meters, D-GIWL.

(2) P: Dongfang Appoach, ACA606, 5100 meters maintaining, estimating DEO 05, Information G received.

C: ACA606, Dongfang Approach, radar contact, follow DEO03A Arrival, Information G is valid, expect ILS approach, Runway 36.

(3) C: N3929G, turn right heading 090 for base, reduce speed to 220 knots, you are No.1.

P: Right heading 090, 220 knots, No.1, N3929G.

(4) C: B-HCJC, continue present heading due separation, descend to and maintain 1800 meters, you are No.2.

P: Continue present heading, descending to 1800 meters. No.2, B-HCJC.

(5) C: THY50, Dongfang Approach, after BRO, fly heading 180 for downwind.

P: After BRO, heading 180 for downwind, THY50.

(6) C: G-PSTO, Dongfang Approach, radar contact, follow ESP01A arrival, expect ILS approach Runway 36.

P: ESP01A arrival, expect ILS approach Runway 36, G-PSTO.

(7) C: UPS390 contact Dongfang Control 123.35, good day.

P: Contact Dongfang Control 123.35, good day, UPS390.

(8) C: FIN132, Dongfang Approach, resume own navigation, turn left direct to CHM.

P: Resume own navigation, turning left direct to CHM, FIN132.

答案:

航空器 a: N3929G, 航空器 b: B-HCJC, 航空器 c: ACA606,

航空器 d: G-PSTO, 航空器 e: THY50, 航空器 f: D-GIWL,

航空器 g: FIN132, 航空器 h: UPS390.

2.

答案:

(1) N3929G, Dongfang Approach, turn right heading 090 for base.

(2) B-HCJC, Dongfang Appraoch, continue present heading, extend downwind for spacing.

（3）ACA606, Dongfang Approach, radar contact, maintain present level, follow DEO11 arrival, Information S is valid.

（4）G-PSTO, Dongfang Approach, radar contact, maintain present level, expect vectoring for ILS approach, Runway 36, Information Q is valid.

（5）THY50, Dongfang Approach, radar contact, maintain present level, follow BRO11 arrival, expect ILS approach Runway 36, Information W.

（6）D-GIWL, Dongfang Approach, radar contact, continue runway heading, climb to and maintain 4,500 meters on standard.

（7）FIN132, Dongfang Approach, resume own navigation, turn left direct to CHM.

（8）UPS390, Dongfang Approach, contact Dongfang Control on 120.36, good day.

四、思考题

答案：

C: DGIWL, climb to and maintain 2700 meters.

C: THY50, descend to and maintain 3000 meters on QNH 1011.

五、故事复述

录音原文：

During early September 2000, a Pro Air 737-400 aircraft that was flying at 10,000 feet and 314 knots near LaGuardia Airport in New York sustained significant windshield damage from a bird strike. The strike occurred during daytime visual meteorological conditions. Glass fragments from the windshield caused minor lacerations on both of the captain's arms. The captain also reported that his face felt "sunburned". Fortunately, the captain was not severely injured and the damaged windshield did not lead to a loss of cabin pressure. Both the captain and the first officer were able to perform their duties throughout this event and were able to make an uneventful landing.

六、小拓展：不守规矩的乘客（unruly passengers）

（1）unruly; violent; aggressive; divert; arrest; disturbance; disruptive; threatening; security

（2）第一次：We've got a situation with a passenger. We're gonna have to return back to the airport. Right now it's not an emergency but we'd like you to know if it turns into one.

第二次：UAL1074 is declaring an emergency due to a passenger disturbance. He's restrained. We need to return to the airport.

第三次：We have a passenger becoming violent. No weapons involved. He's restrained by other passengers now though. We don't know his mental condition but sounds like he's restrained for now.　We just need to get on the ground.

第四次：There's a passenger in the back. He ran forward towards the cockpit and he is being restrained by other passengers. The cockpit is secure and we would just expect to return to the airport and have the authorities meet him.

（3）What's the nature of the emergency?

（4）Say disturbance level.

（5）Is the passenger restrained?

第6章　机场管制——最后进近及着陆阶段

6.1　起落航线飞行

1. 听力练习

录音原文及答案：

（1）**C:** CHB6263, extend downwind, No. 2, follow B737 on final.

　　　P: No. 2, B737 in sight, CHB6263.

（2）**C:** CSS467, make one orbit right due traffic on the runway, report again on final.

　　　P: Orbit right, CSS467.

（3）**C:** CBJ5323, No.1, make short approach, B737 6 miles final.

　　　P: Short approach, CBJ5323.

（4）**C:** B-7921, extend downwind, No. 2, follow the light aircraft on 4 kilometers final.

　　　P: Extend downwind, B-7921.

（5）**C:** CSN3356, join right hand downwind, Runway 36, wind 170 degrees, 15 m/s, QNH 1002.

　　　P: Join right hand downwind, Runway 36, QNH 1002, CSN3356.

2. 听力练习

录音原文：

（1）Dongfang Tower, AAR415, turning final, Runway 36.

（2）FIN8721 is about to turn downwind, Runway 36.

（3）GIA169 is about to join base, Runway 36.

（4）BAW4140 has just turned downwind, Runway 36.

（5）UAE756, Runway 36, on runway heading, climbing to 600 meters.

答案：

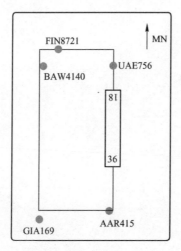

图 6-1

3. 听说练习

录音原文:

（1）Dialogue 1

P: Dongfang Tower, G-LCDB, Cessna172, 12 kilometers north of the field, 900 meters on QNH 1001, Information I, request joining instructions to land.

C: G-LCDB, Dongfang Tower, join right hand downwind, Runway 36, Information I QNH 1001 is valid.

P: Right hand downwind, Runway 36, G-LCDB.

（2）Dialogue 2

P: Dongfang Tower, YV-OFH, Cessna172, 15 kilometers south of the field, 900 meters, request joining instructions to land.

C: YV-OFH, Dongfang Tower, make straight-in approach, Runway 36, wind 190 degrees, 14 m/s, QNH 1001.

P: Straight-in approach, Runway 36, QNH 1001, YV-OFH.

（3）Dialogue 3

P: Dongfang Tower, D-AHGO, downwind.

C: D-GO, No. 2, follow the MD-11 on base.

P: No.2, traffic in sight, D-GO.

(A moment later)

C: D-GO, wind 190 degrees, 15 m/s, Runway 36, cleared to land.

P: Runway 36, cleared to land, D-GO.

答案:

（1）Dialogue 1

① G-LCDB.

② north.

③ right hand downwind.

（2）Dialogue 2

① YV-OFH.

② straight-in approach.

③ wind 190 degrees, 14 m/s.

（3）Dialogue 3

① D-AHGO.

② downwind.

③ MD-11, on base.

6.2　最后进近与着陆

2. 听力练习

录音原文及答案:

（1）**P:** Dongfang Tower, CSH9129, outer marker.

　　C: <u>CSH9129, Dongfang Tower, report short final</u>.

　　P: Wilco, CSH9129.

　　　　(A moment later)

　　P: CSH9129, short final.

　　C: <u>CSH9129, wind 340 degrees, 7 m/s, Runway 36L, cleared to land</u>.

　　P: Runway 36L, cleared to land, CSH9129.

（2）**P:** Dongfang Tower, CSN3356, final.

　　C: CSN3356, Dongfang Tower, you are No.1 to land. <u>Caution wind shear reported at 600 feet 2 miles final</u>.

　　P: No.1 to land, roger, CSN3356.

　　　　(A moment later)

　　P: CSN3356, going around.

　　C: CSN3356, <u>standard procedure</u>, when passing 1000 feet, turn right to JUI VOR.

（3）**C:** CES5323, <u>the fog is getting dense, RVR is less than 1500 meters, advise go around</u>.

　　P: CES5323, we'll see at our <u>minima</u>.

　　　　(A moment later)

　　P: CES5323, going around.

　　C: CES5323, <u>standard procedure, contact 123.1</u>.

　　P: CES5323, what is the standard procedure?

　　C: <u>CES5323, climb straight ahead to 900 meters and contact 123.1</u>.

　　P: Roger, 900 meters, contact 123.1, CES5323.

3. 口语练习

答案：

图 6-2: Climb straight ahead to 300 meters, turn right, cross TAO at 1200 meters.

图 6-3: Climb straight ahead to 300 meters, turn left, cross PA at 1200 meters.

6.4　本场训练

4. 听力练习

录音原文及答案：

（1）**P:** Dongfang Tower, CKK206, request touch and go.

　　C: CKK206, Dongfang Tower, <u>unable to approve due traffic congestion, make full stop, cleared to land</u>.

　　P: Cleared to land for full stop, CKK206.

（2）**P:** Dongfang Tower, G-CHDG, request touch and go.

　　C: G-CHDG, Dongfang Tower, <u>Runway 36, cleared touch and go</u>.

　　P: Runway 36, cleared touch and go, G-CHDG.

（3）**P:** BAW259, outer marker, request touch and go for training.

C: BAW259, <u>cleared touch and go, make another circuit and report downwind</u>.

P: Roger, cleared touch and go, will call you downwind, BAW259.

C: BAW259, <u>go around immediately, I say again, go around immediately, unidentified vehicle crossing runway</u>.

P: Going around, BAW259.

（4）P: Dongfang Tower, CSN5013, request low pass, <u>unsafe left gear indication</u>.

C: CSN5013, Dongfang Tower, <u>cleared low pass, Runway 27, not below 150 meters, report final</u>.

P: Runway 27, not below 150 meters, CSN5013.

(A moment later)

C: CSN5013, <u>landing gear appears down</u>.

P: Roger, will advise intentions.

（5）P: Dongfang Tower, GCR7843, request low approach, Runway 09 for training.

C: GCR7843, Dongfang Tower, <u>cleared low approach, Runway 09, not below 150 meters, report final</u>.

P: Runway 09, not below 150 meters, report final, GCR7843.

本章综合练习

一、词汇练习

答案：

●●	●●●	●●●	●●●●	●●●●
holding traffic	terminal negative	intention direction	expediting destination	identified

二、词汇练习

答案：

/id/	/t/	/d/
landed taxied vacated awaited instructed	reduced missed	turned confused realized

三、词汇练习

答案：

braking action.

leading edge.

parking bay.

warning light.

boarding gate.

outbound flights.

ground handling personnel.

holding point.

五、完成通话

答案：

（1）**C**: CSN3520, Dongfang Tower, wind 210 degrees at 4 m/s, Runway 22, cleared to land.

（2）**C**: CSN3520, confirm altitude, mode C shows you're too high.

（3）**C**: CSN3520, climb to 1200 meters, fly heading 250, contact Departure on 124.35.

（4）**C**: HVN105, vacate runway at the first convenient right, contact Ground 121.6.

（5）**C**: JAL635, taxi via Taxiway C,C1 to Stand D3.

六、故事复述

录音原文：

It happened just after 2 o'clock in the morning in 1999 at Chicago O'Hare International Airport. Air China had just landed and was rolling out on Runway 14R when the controller instructed Korean Air to taxi into position and hold. After Air China exited the runway at Taxiway T10, the controller instructed the aircraft to turn left onto Taxiway K and cross Runway 27L. The controller then cleared Korean Air for take-off. As the aircraft was rolling down the runway, Air China deviated from its assigned taxi route and entered Runway 14R. The Korean Air captain saw the 747 taxiing onto the runway, but it was too late to stop. Instead, Korean Air 36 lifted off earlier than normal and banked left to avoid striking Air China. The two aircraft missed colliding by about 80 feet.

七、小拓展：如何在无线电陆空通话中更好地沟通？

（1）You want a vector to Virginia Highlands, is that correct?

Are you declaring an emergency?

（2）What assistance do you require? What's the nature of the emergency? What are your intentions?

（3）Virginia Highlands airport right now is about 5 miles east - northeast of your position. Do you want to continue towards Virginia Highlands or do you see somewhere that you might be able to put it down?

（4）Interstate 81 is about, I-81 is about 2, maybe a mile and a half southeast of your position.

There are obstructions about - the first one is about a mile southwest of your position at 2000 ft.

There are multiple obstructions to the southwest extending to the south of your position, the highest one is about 2900 ft, and the lowest is 2000 ft.

If you can still hear me, there is another obstruction about a mile and a half southeast of Virginia Highlands, 2600 ft.

（5）This is N721AT. I am relaying for 45X. He's put it down on 81 and he sees Exit 10 sign.

（6）N721AT, that's excellent. Thank you very much.I'm gonna hold on to you for just a little bit longer, then switch you to Indianapolis Center. If for some reason we lose Comms, the next frequency will be 126.57.

第7章 紧 急 情 况

7.1 发布与认收

1. 听力练习

录音原文：

（1）**P:** MAYDAY MAYDAY MAYDAY, Dongfang Approach, F-AACV, fuel emergency, ditching in sea. Position 70 kilometers west of Dongfang, 3000 meters, heading 090, speed 160 knots, 3 persons on board.

（2）**P:** MAYDAY MAYDAY MAYDAY, Dongfang Approach, D-AXFB, MA from Nanping Airport. I'm unsure of my position. Endurance 15 minutes. Request radar vectoring to Dongfang.

（3）**P:** PANPAN, PANPAN, PANPAN, Dongfang Approach, G-CEPT, one passenger severely ill. 30 kilometers south of airfield at 1200 meters, heading 340. Request priority landing and medical assistance.

（4）**P:** PANPAN, PANPAN, PANPAN, Dongfang Approach, AFR187, 87 passengers and 5 crewmembers on board. No.1 engine shutdown, 30 kilometers northwest of the aerodrome, 2700 meters on QNH 1003.

（5）**P:** PANPAN, PANPAN, PANPAN, Dongfang Approach, G-KLXW. We are coming back to Nanping Airport. There seems to be a bomb on board. Position 40 kilometers, east of Nanping Airport, heading 270, 2700 meters. Request priority landing and emergency services.

答案：

（1）Fuel emergency, ditching in sea.

（2）Unsure of the aircraft's position. Request radar vectoring to Dongfang.

（3）One passenger severely ill. Request priority landing and medical assistance.

（4）No. 1 engine shutdown.

（5）There seems to be a bomb on board. Request priority landing and emergency services.

7.2 处置与取消

1. 听力练习

录音原文：

P: Approach, AIC349. We got a fire warning in the cargo hold. Request immediate descent and landing priority.

C: AIC349, roger. Descend initially to 3000 meters on QNH 1007. I'll call you back for further descent.

P: AIC349, descending to 3000 meters on 1007. We are beginning to have smoke in the cabin.

C: AIC349, roger. Fly heading 220. I'll take you to the runway threshold direct.

P: Roger, we might evacuate on the runway after landing, please advise the rescue service.

C: AIC349, fire trucks are coming out for you. They want to know if you have any dangerous goods on board.

P: Affirm. We've got some kind of chemicals in the hold.

C: AIC349, roger, chemicals. We also need your number of persons on board and fuel on board.

P: We have 212 passengers and 8 crewmembers, and 14 thousand pounds of fuel.

C: Roger, 212 plus 8 and 14 thousand pounds of fuel. AIC349, you are cleared to land.

P: Cleared to land, AIC349.

答案：

（1）They got a fire warning in the cargo hold.

（2）Immediate descent and landing priority.

（3）They might evacuate on the runway after landing.

（4）Information about the dangerous goods on board, number of persons on board and fuel on board.

2. 听力训练

录音原文及答案：

（1）**P:** CPA748, we've just had to dive to avoid colliding with converging traffic.

  C: Do you have any other details? Did you see the type or the markings?

  P: It was a white jet, that's all we know.

  C: Do you wish to file an airmiss report?

  P: Affirm. It was a very close thing. I'll check if the passengers are OK.

   （A moment later）

  P: CPA748, 6 passengers have been badly bruised, but there's a doctor on board, so we'll continue on our route.

  C: Roger, CPA748.

（2）**P:** ALK866, we have a serious fuel leak, request divert to Dongfang.

  C: ALK866, turn right now, heading 280, descend to 3600 meters.

  P: Turning right heading 280, descending to 3600 meters, ALK866.

  C: Do you require emergency assistance at Dongfang?

  P: Affirm, ALK866.

  C: Roger, will advise.

（3）**C:** JAL7908, Dongfang Control, your company has informed us that you may have a bomb on board.

  P: Do you have any information about the type of the bomb?

  C: Negative.

  P: Diverting immediately to Nanping Airport, request emergency services on landing, JAL7908.

本章综合练习

一、听力练习

录音原文：

（1）Cathay 179;　　　（2）Cargo King 429;　　　（3）China Southern 686;

（4）Air France 6393;　　　（5）Emirates 4213;　　　（6）Qantas 5797;

（7）Aeroflot 4031;　　　（8）Singapore 7951;　　　（9）Swiss 6833;

（10）Vietnam Airlines 1255.

答案：

（1）c　　　　（2）b　　　　（3）a　　　　（4）b　　　　（5）a

（6）a　　　　（7）b　　　　（8）a　　　　（9）c　　　　（10）a

二、听力练习

（1）

录音原文：

P: AAR311, Shanghai Approach, we've shut down No. 1 engine after a bird strike. We're coming back.

C: AAR311, roger, do you require landing priority?

P: Negative. There is no fire warning, AAR311.

C: AAR311, roger, turn left heading 250.

答案：Because of bird strike /bird ingestion.

（2）

录音原文：

P: AMU030, we're returning. We seem to have a wheel well fire. The warning light has just flashed on. Request priority landing and emergency services.

C: Roger, AMU030, I'll call you back.

　　（A moment later）

C: AMU030, you're No.1 to land, call Tower on 118.5.

P: 118.5, AMU030.

答案：Wheel well fire.

（3）

录音原文：

P: YZR7991, we have an engine failure. We intend to return to Nanping airport, but we have to dump 40 tons of fuel first.

C: Roger, YZR7991, proceed to fuel dumping area, at 4500 meters, right pattern over MIZ. Report when reaching.

P: YZR7991, 4500 meters over MIZ.

　　（A moment later）

P: YZR7991, reaching MIZ, ready to dump fuel.

C: Roger, go ahead YZR7991, break break. All aircraft, Dongfang Control, fuel dumping in progress, DC8, on radial 120 MIZ VOR, ranging 14 to 20 miles, avoid flight below 4500 meters within 10 miles of fuel dumping track.

P: YZR7991, fuel dumping completed, request approach to Nanping airport.

答案：Dump fuel.

（4）

录音原文：

P: HVN3565, request divert to Dongfang, a passenger is seriously ill, probably a heart attack.

C: Roger, HVN3565, turn right heading 290, I'll tell Dongfang you require medical assistance on landing.

P: Turning right 290, HVN3565.

答案：A passenger is seriously ill.

（5）

录音原文：

P: FIN070, we have lost all electrical power, except the emergency circuit. Request to divert immediately to Taiyuan.

C: FIN070, Roger, turn left heading 030, descend to 4500 meters.

P: Turning left heading 030, leaving 7800 meters, descending to 4500 meters, FIN070.

答案：Lost all electrical power except the emergency circuit.

三、听力练习

录音原文：

C1: QTR5844, cancel SID, assigned heading 230, climb to and maintain 1200 meters, cleared for take-off.

P: Cancel SID, heading 230, 1200 meters, cleared for take-off, QTR5844.

　　(A moment later)

P: PANPAN, PANPAN, PANPAN, QTR5844, QTR5844, QTR5844, engine failure, turning right heading 270 for terrain clearance.

C1: QTR5844, roger, when ready advise intentions and the nature of your emergency.

C1: QTR5844, there appears to be flames and smoke coming from your right hand outboard engine.

P: QTR5844, request vectors to a 10 mile final ILS Runway 21 and we'll require emergency services on arrival.

C1: QTR5844, contact Dongfang Approach 123.7. He has been advised of your emergency.

P: 123.7, QTR5844.

P: Dongfang Approach, QTR5844.

C2: QTR5844, maintain 1200 meters, fly heading 020, vector for downwind, Runway 21 ILS.

P: Heading 020, 1200 meters, QTR5844.

C2: QTR5844, when able, go ahead persons on board and dangerous goods.

P: POB 251, nil dangerous goods, QTR5844.

C2: QTR5844, roger, turn right heading 180, cleared for ILS approach Runway 21.

P: Right heading 180, cleared for ILS approach Runway 21, QTR5844.

C2: QTR5844, contact Tower 120.5,good day.

答案：

（1）Engine failure.

（2）Request vectors to a 10 mile final ILS Runway 21 and require emergency services on arrival.

（3）答案如下图所示。

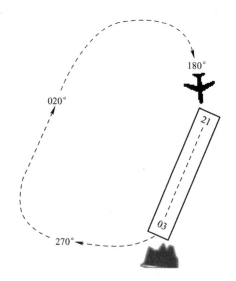

四、口语练习

答案：

Student A

（1）Why is the captain throttling back?

（2）What did Center tell you to do?

（3）Who are you advising that you'll need to hold?

（4）How long is your endurance?

（5）Where is the LAM holding pattern?

（6）What was the reason for a lot of delays?

（7）What was an MD81 instructed to do?

（8）How far is the B737 away from the LAM VOR?

（9）Which direction was the MD-81 heading inbound to LAM?

（10）Where was the MD81 with respect to the B737?

Student B

（1）What is the captain doing?

（2）How long did Center tell us to stay at this altitude?

（3）What are you advising air traffic?

（4）What do you have more than 3 hours?

（5）Where is the VOR that the LAM holding pattern is based on?

（6）What were the results of single-runway operation at Heathrow?

（7）When did the MD81 enter the pattern?

（8）Where did the B737 have approximately 5 miles to go?

（9）What was heading west inbound to LAM?

（10）How far was the MD81 below the B737?

五、故事复述

录音原文：

The story was recorded in a report in Flight International Magazine. In October 1963, an aircraft, Tu124, was on a flight from Estonia to Moscow when a landing gear problem led to a diversion to Leningrad. While holding prior to landing and about 13 miles from Leningrad airport, the aircraft ran out of fuel. The crew had no choice but to land before reaching the airport. Fortunately, there was a river nearby. The crew managed to make a ditch landing on Neva River, where it remained floating on the surface. The aircraft was towed to shore and all 52 occupants survived.

附录 C 国内常见航空公司的代码、话呼及公司名称

三字码	二字码	英文话呼	中文话呼	公司名称
AHK	LD	AIR HONG KONG	香港货运	香港华民航空
AMU	NX	AIR MACAU	澳门	澳门航空
CAL	CI	DYNASTY	华航	台湾中华航空公司
CAO	CA	AIR CHINA FREIGHT	凤凰	中国国际货运航空
CBJ	JD	CAPITAL JET	神鹿	北京首都航空
CCA	CA	AIR CHINA	国际	中国国际航空
CDG	SC	SHANDONG	山东	山东航空
CES	MU	CHINA EASTERN	东方	中国东方航空
CHB	PN	WEST CHINA	西部	中国西部航空
CHH	HU	HAINAN	海南	海南航空
CKK	CK	CARGO KING	货航	中国货运航空
CQH	9C	AIR SPRING	春秋	春秋航空
CQN	OQ	CHONG QING	重庆	重庆航空
CPA	CX	CATHAY	国泰	国泰航空
CSC	3U	SI CHUAN	四川	四川航空
CSH	FM	SHANGHAI AIR	上航	上海航空
CSN	CZ	CHINA SOUTHERN	南方	中国南方航空
CSS	O3	SHUN FENG	顺丰	顺丰航空
CSZ	ZH	SHENZHEN AIR	深圳	深圳航空
CUA	KN	CHINA UNITED	联航	中国联合航空
CXA	MF	XIAMEN AIR	白鹭	厦门航空
CYZ	8Y	CHINA POST	邮政	中国邮政航空
DER	DA	DEER JET	金鹿	金鹿航空
DKH	HO	AIR JUNEYAO	吉祥	吉祥航空
EPA	DZ	DONGHAI AIR	东海	深圳东海航空
EVA	BR	EVA	长荣	台湾长荣航空公司
GCR	GS	BOHAI	渤海	天津航空
GDC	CN	GRAND CHINA	大新华	大新华航空
HBH	NS	HEBEI AIR	河北	河北航空
HDA	KA	DRAGON	港龙	国泰港龙航空
HKE	UO	HONGKONG SHUTTLE	香港	香港快运航空
HXA	G5	CHINA EXPRESS	华夏	华夏航空
JAE	JI	JADE CARGO	翡翠	翡翠国际货运航空

三 字 码	二 字 码	英文话呼	中文话呼	公 司 名 称
JOY	JR	JOY AIR	幸福	幸福航空
KNA	KY	KUNMING AIR	昆航	昆明航空
KPA	VD	KUN PENG	鲲鹏	鲲鹏航空
LKE	8L	LUCKY AIR	祥鹏	祥鹏航空
OKA	BK	OKAY JET	奥凯	奥凯航空
SHQ	F4	SHANGHAI CARGO	上货航	上海国际货运航空
TBA	TV	TIBET	西藏	西藏航空
UEA	EU	UNITED EAGLE	锦绣	成都航空
YZR	Y8	YANGTZE RIVER	扬子江	扬子江快运
CCD	CA	XIANGJIAN	响箭	大连航空
CDC	GJ	HUALONG	华龙	浙江长龙货运航空

附录 D 国外常见航空公司的代码、话呼及公司名称

三 字 码	二 字 码	英 文 话 呼	英 文 名 称	中 文 名 称
AAL	AA	AMERICAN	American Airlines	美国航空
AAR	OZ	ASIANA	Asiana Airlines	韩亚航空
ABW	RU	AIRBRIDGE CARGO	Airbridge Cargo Airlines	俄罗斯空桥货运航空
ACA	AC	AIR CANADA	Air Canada	加拿大航空
AEA	UX	EUROPA	Air Europa	西班牙欧洲航空
AFG	FG	ARIANA	Ariana Afghan Airlines	阿里亚那阿富汗航空
AFL	SU	AEROFLOT	Aeroflot - Russian Airlines	俄罗斯航空
AFR	AF	AIR FRANS	Air France	法国航空
AIC	AI	AIR INDIA	Air India Limited	印度航空
ALK	UL	SRILANKAN	Srilankan Airlines	斯里兰卡航空
ANA	NH	ALL NIPPON	All Nippon Airways	全日空航空
ANZ	NZ	NEW ZEALAND	Air New Zealand	新西兰航空
AUA	OS	AUSTRIAN	Austrian Airlines	奥地利航空
AZA	AZ	ALITALIA	Alitalia Societa Aerea Italiana S.P.A	意大利航空
BAW	BA	SPEEDBIRD	British Airways	英国航空
BKP	PG	BANGKOK AIR	Bangkok Airways	泰国曼谷空
CEB	5J	CEBU AIR	Cebu Pacific Air	菲律宾宿务太平洋航空
CLX	CV	CARGOLUX	Cargolux Airlines International S.A.	卢森堡货运航空
DAL	DL	DELTA	Delta Air Lines	美国达美航空
DLH	LH	LUFTHANSA	Deutsche Lufthansa, AG, Koeln	德国汉莎航空
EIA	EZ	EVERGREEN	Evergreen International Airlines	美国长青航空
ELY	LY	ELAL	EL AL - Israel Airlines	以色列航空
FDX	FX	FEDEX	Federal Express	美国联邦快递
FIN	AY	FINN AIR	Finnair OYJ	芬兰航空
GEC	LH	LUFTHANSA CARGO	Lufthansa Cargo AG, Frankfurt	德国汉莎货运航空
GFA	GF	GULF AIR	Gulf Air	阿联酋海湾航空
GIA	GA	INDONESIA	Garuda Indonesia Airways, PT.	印度尼西亚鹰航空
HVN	VN	VIET NAM AIRLINES	Hang Khong Viet Nam	越南航空
IAW	IW	IRAQI	Iraqi Airways	伊拉克航空
IRA	IR	IRAN AIR	Iran Air	伊朗航空
JAL	JL	JAPAN AIR	Japan Airlines	日本航空
KAC	KU	KUWAITI	Kuwait Airways	科威特航空
KAL	KE	KOREAN AIR	Korean Air Lines	大韩航空

续表

三 字 码	二 字 码	英 文 话 呼	英 文 名 称	中 文 名 称
KLM	KL	KLM	KLM Royal Dutch Airlines	荷兰皇家航空
KOR	JS	AIR KORYO	Air Koryo	朝鲜高丽航空
KZR	KC	ASTANALINE	Air Astana	哈斯克斯坦阿斯塔那航空
MAS	MH	MALAYSIAN	Malaysia Airlines	马来西亚航空
MGL	OM	MONGOL AIR	Mongolian Airlines	蒙古航空
MPH	MP	MARTIN AIR	Martinair Holland N.V.	荷兰马丁航空
MSR	MS	EGYPT AIR	Egypt Air	埃及航空
MXA	MX	MEXICANA	Mexicana Airlines	墨西哥航空
NCA	KZ	NIPPON GARGO	Nippon Cargo Airlines	日本货运航空
OAL	OA	OLYMPIC	Olympic Air	奥林匹克航空
OEA	OX	ORIENT THAI	Orient Thai Airlines	泰国东方航空
PAA	PA	CLIPPER	Pan American Airways	美国泛美航空
PAC	PO	POLAR	Polar Air Cargo Worldwide	美国博立航空
PAL	PR	PHILIPPINE	Philippine Air Lines	菲律宾航空
PIA	PK	PAKISTAN	Pakistan International Airlines	巴基斯坦国际航空
QFA	QF	QANTAS	Qantas Airways	澳大利亚快达航空
QTR	QR	QATARI	Qatar Airways	卡塔尔航空
RBA	BI	BRUNEI	Royal Brunei Airlines	文莱皇家航空
RNA	RA	ROYAL NEPAL	Royal Nepal Airlines	尼泊尔皇家航空
SAA	SA	SPRINGBOK	South African Airways	南非航空
SAS	SK	SCANDINAVIAN	Scandinavian Airlines System	北欧航空
SBI	S7	SIBERIAN AIRLINES	Siberia Airlines	俄罗斯西伯利亚航空
SIA	SQ	SINGAPORE	Singapore Airlines	新加坡航空
SLK	MI	SILKAIR	Silkair	新加坡胜安航空
SVA	SV	SAUDIA	Saudi Arabian Airlines	沙特阿拉伯航空
SWR	LX	SWISS	Swiss International Air Lines	瑞士国际航空
TAY	3V	QUALITY	TNT Airways	比利时 TNT 航空
THA	TG	THAI	Thai Airways International	泰国国际航空
THY	TK	TURKISH	Turkish Airlines	土耳其航空
UAE	EK	EMIRATES	Emirates	阿联酋航空
UAL	UA	UNITED	United Airlines	美国联合航空
UKR	6U	AIR UKRAINE	Air Ukraine	乌克兰航空
UPS	5X	UPS	United Parcel Service	美国联合包裹航空
UZB	HY	UZBEK	Uzbekistan Airways	乌兹别克斯坦航空
VIR	VS	VIRGIN	Virgin Atlantic	英国维珍航空

参 考 文 献

[1]   ICAO. Manual of radiotelephony[S]. 4th ed. Montreal：International Civil Aviation Organization，2007.

[2]   中国民用航空局. 空中交通无线电通话用语指南[M]. 成都：西南交通大学出版社，2005.

[3]   杜实. 空中交通监视服务[M]. 北京：中国民航出版社，2012.

[4]   闫少华. 非常规陆空通话英语[M]. 北京：中国民航出版社，2008.

[5]   刘继新. 特殊情况下的无线电通话用语[M]. 北京：国防工业出版社，2010.

[6]   ICAO. Manualon the implementation of ICAO language proficiency requirements[S]. 2nd ed. Montreal：International Civil Aviation Organization，2010.

[7]   ICAO. Aeronautical telecommunications[S]. 7th ed. Montreal：International Civil Aviation Organization，2016.

[8]   ICAO. Air traffic management[S]. 16th ed. Montreal：International Civil Aviation Organization，2016.

[9]   FAA. Contractions [S]. Washington DC：Federal Aviation Administration，2021.